AF247471

STOCHASTIC ANALYSIS AND APPLICATIONS
VOLUME 2

PREFACE

On September 1~5, 2000, the Sixth International Conference on Nonlinear Functional Analysis and Applications was held at Gyeongsang National University (GNU) and Kyungnam University (KU) by the financial support of the Institutes of Basic Science Research of GNU and KU.

The aim of the conference is to introduce and exchange recent new topics on the areas of fixed point theory, variational inequality and complementarity problem theory, nonlinear ergodic theory, difference, differential and integral equations, control and optimization theory, dynamic system theory, inequality theory, stochastic analysis and probability theory, and their applications. Especially, 35 invited talks were delivered by leading experts in these areas and 100 more talks also were given by many scholars including 45 scholars from overseas at the conference.

We have published the Proceedings of the Conference every year and so we will publish four Volumes of the Proceedings 2000 of the Sixth International Conference on Nonlinear Functional Analysis and Applications as follows:

(∗) Fixed Point Theory and Applications, Vol. 2(2001), Vol. 3(2001)
    (Co-Editors: Yeol Je Cho, Jong Kyu Kim, Shin Min Kang),

(∗) Differential Equations and Applications, Vol. 2(2001)
    (Co-Editors: Yeol Je Cho, Jong Kyu Kim, Ki Sik Ha),

(∗) Inequality Theory and Applications, Vol. 1(2001)
    (Co-Editors: Yeol Je Cho, Jong Kyu Kim, Sever S. Dragomir),

(∗) Stochastic Analysis and Applications, Vol. 2(2001)
    (Co-Editors: Yeol Je Cho, Jong Kyu Kim, Yong Kab Choi).

Each Volume of the Proceedings 2000 contains the original papers on the topics mentioned above, which were delivered at the conference and reviewed by the referees and the members of the Editorial Board for the Proceedings, and some invited papers of leading scholars in their fields as special contributors.

We would like to express our deep appreciation to all talkers and participants of the conference for their concern and encouragement. Especially, we are very grateful to Professor H. Tanabe, Otemon Gakuin University, Japan, Professor S. S. Chang, Sichuan University, People's Republic of China, Professor K. S. Ha, Professor J. Y. Park and Professor Y. H. Lee, Pusan National University, Korea, Professor S. Park and Professor D. Kim, Seoul National University, Korea, Professor D. M. Jung, Sogang University, Korea, Professor Z. Lin, Zhejiang

University, People's Republic of China, Professor Ronald E. Bruck, University of Southern California, U. S. A., Professor Jean B. Baillon, Universite Lyon, France, Professor B. D. Rouhani, Shahid Beheshti University, Iran, Anthony T. M. Lau, University of Alberta, Canada, Professor Q. M. Shao, University of Oregon, U. S. A., Professor X. P. Ding, Sichuan Normal University, People's Republic of China, Professor T. S. Kim, Wonkwang University, Korea, Professor U J. Choi and Dr. P. S. Kim, KAIST, Korea, Professor D. H. Pahk, Yonsei University, Korea, Professor W. S. Kim, Hanyang University, Korea, Professor K. W. Jun and Professor C. K. Park, Chungnam National University, Korea, Professor Q. H. Choi, Inha University, Korea, Professor T. I. Jeon, Taejon University, Korea, Professor W. K. Kim, Chungbuk National University, Korea, Professor H. Y. Zhou, Shijiazhuang Mechanical Engineering College, People's Republic of China, Professor Alexander G, Ramm, Kansas State University, U. S. A., Professor C. E. M. Pearce, University of Adelaide, Australia, Professor S. S. Dragomir, Victoria University of Technology, Australia, Professor W. Takshashi, Tokyo Institute of Technology, Japan, Professor N. Hirano and Professor K. Nishiura, Yokohama National University, Japan, Professor L. S. Liu, Qufu Normal University, People's Republic of China, Professor F. Zhang, Xuzhou Normal University, People's Republic of China, Professor A. Yagi, Osaka University, Japan, Professor K. Balanchandran, Professor S. M. Anthoni and Professor R. Sakthivel, Bharathiar University, India, Professor S. C. Palaniswami, Kongku Nadu Arts and Science College, India, Professor Z. Liu, Liaoning Normal University, People's Republic of China, Professor G. Li, Yangzhou University, People's Republic of China, Dr. George X. Z. Yuan, University of Queensland, Australia, Professor D. S. Kim, Professor G. M. Lee, Professor T. H. Kim and J. M. Jeong, Pukyong National University, Korea, Professor B. S. Lee, Professor H. Y. Lee and I. Y. Lee, Kungsung University, Korea, Professor Y. H. Lee, K. B. Park, Y. M. Nam, Kyungnam University, Korea, Professor J. S. Ume, Changwon National University, Korea, Professor R. Baik, Honam University, Korea, Professor J. Choi, Dongguk University, Korea, Professor J. S. Bae, Sungkyunkwan University, Korea, Professor S. S. Kim, Dongeui University, Korea, Professor S. H. Kum, Korea Maritime University, Korea, Professor C. J. Zhang, People's Republic of China, Professor J. Zhu, Xuzhou Normal University, People's Republic of China, and many others for giving their invited talks at the conference and their encouragement in organizing the conference.

Further, we are also very grateful to Professor B. E. Rhoades, Indiana University, U. S. A., Professor Ravi P. Agarwal, National University of Singapore, Singapore, Professor George A. Anastassiou, University of Memphis, U. S. A., Professor S. Reich, The Technion-Israel Institute of Technology, Israel, Professor B. Fisher, University of Leicester, England, Professor Paul C. S. Lin, Bishops University, Canada, Professor D. O'Regan, National University of Ireland, Ireland, Professor M. O. Osilike, University of Nigeria, Nigeria, Professor L. J. Lin, National Changhua University of Education, Republic of China, Professor M. A. Noor, Dalhousie University, Canada, Professor N. S. Barnett, Professor P. Cerone, Professor J. Roumeliotis and Professor A. Sofa, Victoria University of Technology, Australia, Professor J. Pečarić and Dr. I. Budimir, University of Zagreb, Croatia, Professor M. Matić, University of Split, Croatia, Professor Y. Q. Chen and Professor N. J. Huang, Sichuan University, People's Republic of China, Professor V. N. Phat and Professor N. D. Yen, Hanoi Institute of Mathematics, Vietnam, Professor W. L. Yang, Armored Force Engineering Institute, People's Republic of China, Professor C. Buşe, Professor V. Radu and Professor L. Radu, West University of Timisoara, Romania, Professor Ş. M. Şoltuz, Wilhelm Raabe Str. Nr. 26, AP. 1408, Germany, Professor S. Cobazaş and Professor R. Precup, Babeş-Bolyai University, Romania, Professor V. Gluščević, Royal Australian Air Force, Australia, Professor A. Lupaş and Professor L. Lupaş, Universitatea Româno-Germană din Sibiu, Romania, Professor C. P. Niculescu University of Craiova, Romania, Professor F. Qi, Jiaozuo Institute of Technology, People's Republic of China, Professor E. M. Sztendur Victoria University of Technology, Australia, Professor Ž. Tomovski, Faculty of Mathematical and Natural Sciences, Macedonia, Professor Y. F. Su, Cangzhou Normal College, People's Republic of China, Professor L. X. Zhang, Zhejiang University, People's Republic of China, Professor Y. H. Wang, Tunghai University, Taiwan, Professor R. U. Verma, University of Toledo, U. S. A., Professor X. Wu, Yunnan Normal University, People's Republic of China, for sending their new papers as special contributors for the Proceedings 2000.

Also we give our deep thanks to all faculty members and all graduate students of Departments of Mathematics and Mathematics Education, Gyeongsang National University and Kyungnam University for their concern and help and, further, typing and arranging all the papers submitted to the Proceedings 2000.

Finally, we thank Dr. Frank Columbus, President and Editor-in-Chief, and Dr. Nadya S. Columbus, Vice-President, and all members of Nova Science Publishers, Inc., New York, U. S. A., for their pleasant cooperation and publishing our Proceedings every year.

Finally, also the 7th International Conference on Nonlinear Functional Analysis and Applications will be held at GNU and KU on August 6~10, 2001. We are looling forward to seeing you at our Conference 2001 in Chinju and Masan, Korea.

July 30, 2001

Co-Editors: Yeol Je Cho, Jong Kyu Kim, Yong Kab Choi

APPROXIMATING CSISZÁR f-DIVERGENCE VIA TWO INTEGRAL IDENTITIES AND APPLICATIONS

N. S. BARNETT[1], P. CERONE[1], S. S. DRAGOMIR[1] AND J. ROUMELIOTIS[1]

ABSTRACT. Some approximations of the Csiszár f-divergence via the use of the integral identities obtained in [8] and [9] and applications are given.

1. Introduction

One of the important issues in many applications of Probability Theory is finding an appropriate measure of *distance* (or *difference* or *discrimination*) between two probability distributions. A number of divergence measures for this purpose have been proposed and extensively studied by Jeffreys [23], Kullback and Leibler [32], Rényi [42], Havrda and Charvat [21], Kapur [26], Sharma and Mittal [15], Burbea and Rao [5], Rao [41], Lin [34], Csiszár [14], Ali and Silvey [1], Vajda [51], Shioya and Da-te [45] and others (see for example [26] and the references therein).

These measures have been applied in a variety of fields such as: anthropology [37], genetics [39], finance, economics, and political science [47], [48], [43], biology [39], the analysis of contingency tables [20], approximation of probability distributions [27], [24], signal processing [25], [3] and pattern recognition [10], [53]. A number of these measures of distance are specific cases of Csiszár f-divergence and so further exploration of this concept will have a flow on effect to other measures of distance and to areas in which they are applied.

Assume that a set χ and the σ-finite measure μ are given. Consider the set of all probability densities on μ to be $\Omega := \left\{ p \mid p : \chi \to \mathbb{R}, \ p(x) \geq \right.$

Received March 14, 2001.

2000 Mathematics Subject Classification. Primary 94A99; Secondary 26D15.

Key Words and Phrases. Csiszár f-divergence, analytic inequalities.

[1] School of Communications and Informatics, Victoria University of Technology, PO Box 14428, Melbourne, MC 8001, Australia. Email addresses: neil@matilda.vu.edu.au http://sci.vu.edu.au/staff/neilb.html, pc@matilda.vu.edu.au http://sci.vu.edu.au/staff/peterc.html, sever@matilda.vu.edu.au http:// rgmia.vu.edu.au/SSDragomirWeb.html and john.roumeliotis@vu.edu.au http://dingo.vu.edu.au/126johnr/

$0,\ \int_{\chi} p(x)\, d\mu(x) = 1\}$. The Kullback-Leibler divergence [32] is well known among the information divergences. It is defined as:

$$(1.1) \qquad D_{KL}(p,q) := \int_{\chi} p(x) \log\left[\frac{p(x)}{q(x)}\right] d\mu(x), \quad p,q \in \Omega,$$

where log is to base 2.

In Information Theory and Statistics, various divergences are applied in addition to the Kullback-Leibler divergence. These are the: *variation distance* D_v, *Hellinger distance* D_H [22], χ^2-*divergence* D_{χ^2}, α-*divergence* D_α, *Bhattacharyya distance* D_B [4], *Harmonic distance* D_{Ha}, *Jeffreys distance* D_J [23], *triangular discrimination* D_Δ [49], etc. They are defined as follows, respectively:

$$(1.2) \qquad D_v(p,q) := \int_{\chi} |p(x) - q(x)|\, d\mu(x), \quad p,q \in \Omega,$$

$$(1.3) \qquad D_H(p,q) := \int_{\chi} \left[\sqrt{p(x)} - \sqrt{q(x)}\right]^2 d\mu(x), \quad p,q \in \Omega,$$

$$(1.4) \qquad D_{\chi^2}(p,q) := \int_{\chi} p(x) \left[\left(\frac{q(x)}{p(x)}\right)^2 - 1\right] d\mu(x), \quad p,q \in \Omega,$$

$$(1.5) \quad D_\alpha(p,q) := \frac{4}{1-\alpha^2}\left[1 - \int_{\chi} [p(x)]^{\frac{1-\alpha}{2}} [q(x)]^{\frac{1+\alpha}{2}}\, d\mu(x)\right], \quad p,q \in \Omega,$$

$$(1.6) \qquad D_B(p,q) := \int_{\chi} \sqrt{p(x)\,q(x)}\, d\mu(x), \quad p,q \in \Omega,$$

$$(1.7) \qquad D_{Ha}(p,q) := \int_{\chi} \frac{2p(x)\,q(x)}{p(x) + q(x)}\, d\mu(x), \quad p,q \in \Omega,$$

$$(1.8) \qquad D_J(p,q) := \int_{\chi} [p(x) - q(x)] \ln\left[\frac{p(x)}{q(x)}\right] d\mu(x), \quad p,q \in \Omega,$$

$$(1.9) \qquad D_\Delta(p,q) := \int_{\chi} \frac{[p(x) - q(x)]^2}{p(x) + q(x)}\, d\mu(x), \quad p,q \in \Omega.$$

For other divergence measures, see the paper [26] by Kapur or the book on line [46] by Taneja. For a comprehensive collection of preprints available on line, see the RGMIA web site *http://rgmia.vu.edu.au/papersinfth.html*

Csiszár f-divergence is defined as follows [14]:

$$(1.10) \qquad I_f(p,q) := \int_{\chi} q(x) f\left[\frac{p(x)}{q(x)}\right] d\mu(x), \quad p,q \in \Omega,$$

where f is convex on $(0, \infty)$. It is assumed that $f(u)$ is zero and strictly convex at $u = 1$. By appropriately defining this convex function, various divergences are derived. All the above distances (1.1)~(1.9), are particular instances of Csiszár f-divergence. There are also many others which are not in this class (see for example [26] or [46]). For the basic properties of Csiszár f-divergence see [12]-[16].

The main aim of this paper is to point out some representations of Csiszár f-divergence for the function which has the $(n-1)$-derivative $(n \geq 1)$ absolutely continuous by employing two recent integral identities from [8] and [9] involving interior point and end point identities. Estimates for the remainder are also provided.

2. Representation of Csiszár f-Divergence

In [8] (see also [6]), the authors proved the following integral identity generalising the mid-point rule.

Lemma 1. *Let $g : [a,b] \to \mathbb{R}$ be a function such that $g^{(n-1)}$ is absolutely continuous. Then for all $x \in [a,b]$, we have the identity:*

$$(2.1) \quad \int_a^b g(t)\,dt = \sum_{k=0}^{n-1} \frac{1}{(k+1)!}[(b-x)^{k+1} + (-1)^k(x-a)^{k+1}]g^{(k)}(x)$$
$$+ (-1)^n \int_a^b K_n(x,t)\,g^{(n)}(t)\,dt,$$

where the kernel $K_n : [a,b]^2 \to \mathbb{R}$ is given by

$$(2.2) \quad K_n(x,t) := \begin{cases} \frac{(t-a)^n}{n!}, & a \leq t \leq x \leq b, \\ \frac{(t-b)^n}{n!}, & a \leq x < t \leq b. \end{cases}$$

In particular, if $x = \frac{a+b}{2}$, then

$$(2.3) \quad \int_a^b g(t)\,dt = \sum_{k=0}^{n-1} \frac{1}{2^{k+1}}\left[\frac{1+(-1)^k}{(k+1)!}\right](b-a)^{k+1}\,g^{(k)}\left(\frac{a+b}{2}\right)$$
$$+ (-1)^n \int_a^b M_n(t)\,g^{(n)}(t)\,dt,$$

where

$$(2.4) \quad M_n(t) := \begin{cases} \frac{(t-a)^n}{n!}, & a \leq t \leq \frac{a+b}{2}, \\ \frac{(t-b)^n}{n!}, & \frac{a+b}{2} < t \leq b. \end{cases}$$

Another integral identity generalising the trapezoid rule is embodied in the following lemma (see [9] or [7]).

Lemma 2. *Let $g : [a,b] \to \mathbb{R}$ be as in Lemma 1. Then, for all $x \in [a,b]$, we have the representation*

$$
(2.5) \quad \int_a^b g(t)\, dt = \sum_{k=0}^{n-1} \frac{1}{(k+1)!} \left[(x-a)^{k+1} g^{(k)}(a) + (-1)^k (b-x)^{k+1} g^{(k)}(b) \right]
$$
$$
+ \frac{1}{n!} \int_a^b (x-t)^n g^{(n)}(t)\, dt,
$$

In particular, if $x = \frac{a+b}{2}$, then

$$
(2.6) \quad \int_a^b g(t)\, dt = \sum_{k=0}^{n-1} \frac{1}{2^{k+1}(k+1)!} (b-a)^{k+1} \left[g^{(k)}(a) + (-1)^k g^{(k)}(b) \right]
$$
$$
+ \frac{(-1)^n}{n!} \int_a^b \left(t - \frac{a+b}{2} \right)^n g^{(n)}(t)\, dt.
$$

Let us consider $x = (1-\lambda)a + \lambda b$, $\lambda \in [0,1]$. Then, from (2.1), we obtain

$$
\int_a^b g(t)\, dt
$$
$$
(2.7) \quad = \sum_{k=0}^{n-1} \frac{1}{(k+1)!} \left[(1-\lambda)^{k+1} + (-1)^k \lambda^{k+1} \right] (b-a)^{k+1} g^{(k)}((1-\lambda)a + \lambda b)
$$
$$
+ (-1)^n \int_a^b K_n((1-\lambda)a + \lambda b, t)\, g^{(n)}(t)\, dt,
$$

and, from (2.5), we obtain

$$
\int_a^b g(t)\, dt
$$
$$
(2.8) \quad = \sum_{k=0}^{n-1} \frac{1}{(k+1)!} \left[\lambda^{k+1} g^{(k)}(a) + (-1)^k (1-\lambda)^{k+1} g^{(k)}(b) \right] (b-a)^{k+1}
$$
$$
+ \frac{1}{n!} \int_a^b [(1-\lambda)a + \lambda b - t]^n g^{(n)}(t)\, dt.
$$

We are now able to state and prove the following representation result for the Csiszár f-divergence.

Theorem 1. *Let $f : \mathbb{R} \to \mathbb{R}$ be a function such that $f^{(n)}$ is absolutely continuous on any $[a, b] \subset \mathbb{R}$. If $p, q \in \Omega$, then*

(2.9)
$$
\begin{aligned}
I_f\,(p, q) \\
= f\,(1) + \sum_{k=0}^{n-1} \frac{1}{(k+1)!} \left[(1 - \lambda)^{k+1} + (-1)^k \lambda^{k+1} \right] \\
\times I_{(\cdot-1)^{k+1} f^{(k+1)} [(1-\lambda)+\lambda\cdot]}\,(p, q) + (-1)^n \int_\Gamma q\,(x) \\
\times \left(\int_1^{\frac{p(x)}{q(x)}} K_n \left[\frac{(1 - \lambda)\,q\,(x) + \lambda p\,(x)}{q\,(x)}, t \right] f^{(n+1)}\,(t)\,dt \right) d\mu\,(x)
\end{aligned}
$$

for all $\lambda \in [0, 1]$ and

(2.10)
$$
\begin{aligned}
I_f\,(p, q) \\
= f\,(1) + \sum_{k=0}^{n-1} \frac{\lambda^{k+1}}{(k+1)!} f^{(k+1)}\,(1)\,D_k\,(p, q) \\
+ \sum_{k=0}^{n-1} \frac{(-1)^k (1 - \lambda)^{k+1}}{(k+1)!} I_{(\cdot-1)^{k+1} f^{(k+1)}(\cdot)}\,(p, q) + \frac{1}{n!} \int_\Gamma [q\,(x)]^{-n+1} \\
\times \left(\int_1^{\frac{p(x)}{q(x)}} [\lambda p\,(x) + [(1 - \lambda) - t]\,q\,(x)]^n\, f^{(n+1)}\,(t)\,dt \right) d\mu\,(x),
\end{aligned}
$$

where

$$
D_k\,(p, q) = \int_\Gamma [p\,(x) - q\,(x)]^k\, [q\,(x)]^{-k+1}\, d\mu\,(x).
$$

Proof. If we apply the identity (2.7) for f', we get

(2.11)
$$
\begin{aligned}
f\,(b) = f\,(a) + \sum_{k=0}^{n-1} \frac{1}{(k+1)!} \left[(1 - \lambda)^{k+1} + (-1)^k \lambda^{k+1} \right] \\
\times (b - a)^{k+1}\, f^{(k+1)}\,((1 - \lambda)\,a + \lambda b) \\
+ (-1)^n \int_a^b K_n\,[(1 - \lambda)\,a + \lambda b, t]\, f^{(n+1)}\,(t)\,dt.
\end{aligned}
$$

If, in (2.11), we choose $b = \frac{p(x)}{q(x)}$, $x \in \Gamma$ and $a = 1$, we get

$$
\begin{aligned}
f\left(\frac{p(x)}{q(x)}\right) \\
= f(1) + \sum_{k=0}^{n-1} \frac{1}{(k+1)!} \left[(1-\lambda)^{k+1} + (-1)^k \lambda^{k+1}\right] \\
\times \frac{(p(x) - q(x))^{k+1}}{[q(x)]^{k+1}} \cdot f^{(k+1)} \left[\frac{(1-\lambda)q(x) + \lambda p(x)}{q(x)}\right] \\
+ (-1)^n \int_1^{\frac{p(x)}{q(x)}} K_n \left[\frac{(1-\lambda)q(x) + \lambda p(x)}{q(x)}, t\right] f^{(n+1)}(t)\, dt
\end{aligned}
$$

(2.12)

for all $x \in \Gamma$. If we multiply (2.12) by $q(x) \geq 0$ ($x \in \Gamma$), integrate on Γ and take into account that $\int_\Gamma q(x)\, d\mu(x) = 1$, then we get the representation (2.9). If we apply the identity (2.8) for f', we get

$$
\begin{aligned}
f(b) = f(a) + \sum_{k=0}^{n-1} \frac{1}{(k+1)!} \left[\lambda^{k+1} f^{(k+1)}(a)\right. \\
\left. + (-1)^k (1-\lambda)^{k+1} f^{(k+1)}(b)\right] (b-a)^{k+1} \\
+ \frac{1}{n!} \int_a^b \left[(1-\lambda)a + \lambda b - t\right]^n f^{(n+1)}(t)\, dt.
\end{aligned}
$$

(2.13)

If, in (2.13), we choose $b = \frac{p(x)}{q(x)}$, $x \in \Gamma$ and $a = 1$, we get

$$
\begin{aligned}
f\left(\frac{p(x)}{q(x)}\right) \\
= f(1) + \sum_{k=0}^{n-1} \frac{\lambda^{k+1}}{(k+1)!} f^{(k+1)}(1) \left(\frac{p(x)}{q(x)} - 1\right)^{k+1} \\
+ \sum_{k=0}^{n-1} \frac{(-1)^k (1-\lambda)^{k+1}}{(k+1)!} f^{(k+1)} \left(\frac{p(x)}{q(x)}\right) \left(\frac{p(x)}{q(x)} - 1\right)^{k+1} \\
+ \frac{1}{n!} \int_1^{\frac{p(x)}{q(x)}} \left[\frac{(1-\lambda)q(x) + \lambda p(x)}{q(x)} - t\right]^n f^{(n+1)}(t)\, dt,
\end{aligned}
$$

(2.14)

for all $x \in \Gamma$. If we multiply (2.14) by $q(x) \geq 0$ ($x \in \Gamma$), integrate on Γ and take into account that $\int_\Gamma q(x)\, d\mu(x) = 1$, we get the representation (2.10). This completes the proof.

Remark 1. If, in (2.9), we choose $\lambda = 0$ or $\lambda = 1$ or $\lambda = \frac{1}{2}$, we get, respectively,

$$
\begin{aligned}
I_f(p,q) &= f(1) + \sum_{k=0}^{n-1} \frac{1}{(k+1)!} f^{(k+1)}(1) D_k(p,q) \\
&\quad + (-1)^n \int_\Gamma q(x) \left(\int_1^{\frac{p(x)}{q(x)}} K_n(1,t) f^{(n+1)}(t)\, dt \right) d\mu(x),
\end{aligned}
\tag{2.15}
$$

$$
\begin{aligned}
I_f(p,q) &= f(1) + \sum_{k=0}^{n-1} \frac{(-1)^k}{(k+1)!} I_{(\cdot-1)^{k+1} f^{(k+1)}(\cdot)}(p,q) \\
&\quad + (-1)^n \int_\Gamma q(x) \left(\int_1^{\frac{p(x)}{q(x)}} K_n\left(\frac{p(x)}{q(x)},t\right) f^{(n+1)}(t)\, dt \right) d\mu(x)
\end{aligned}
\tag{2.16}
$$

and

$$
\begin{aligned}
I_f(p,q) \\
= f(1) &+ \sum_{k=0}^{n-1} \left[\frac{1+(-1)^k}{2^{k+1}(k+1)!} \right] I_{(\cdot-1)^{k+1} f^{(k+1)}\left(\frac{1+\cdot}{2}\right)}(p,q) \\
&+ (-1)^n \int_\Gamma q(x) \left(\int_1^{\frac{p(x)}{q(x)}} K_n\left(\frac{q(x)+p(x)}{2q(x)},t\right) f^{(n+1)}(t)\, dt \right) d\mu(x).
\end{aligned}
\tag{2.17}
$$

Remark 2. If, in (2.10), we choose $\lambda = 0$ or $\lambda = 1$ or $\lambda = \frac{1}{2}$, we get, respectively,

$$
\begin{aligned}
I_f(p,q) &= f(1) + \sum_{k=0}^{n-1} \frac{(-1)^k}{(k+1)!} I_{(\cdot-1)^{k+1} f^{(k+1)}(\cdot)}(p,q) \\
&\quad + \frac{1}{n!} \int_\Gamma q(x) \left(\int_1^{\frac{p(x)}{q(x)}} (1-t)^n f^{(n+1)}(t)\, dt \right) d\mu(x),
\end{aligned}
\tag{2.18}
$$

$$
\begin{aligned}
I_f(p,q) &= f(1) + \sum_{k=0}^{n-1} \frac{f^{(k+1)}(1)}{(k+1)!} D_k(p,q) + \frac{1}{n!} \int_\Gamma [q(x)]^{-n+1} \\
&\quad \times \left(\int_1^{\frac{p(x)}{q(x)}} (p(x) - tq(x))^n f^{(n+1)}(t)\, dt \right) d\mu(x)
\end{aligned}
\tag{2.19}
$$

and

$$I_f\left(p,q\right) = f\left(1\right) + \sum_{k=0}^{n-1} \frac{f^{(k+1)}\left(1\right)}{2^{k+1}\left(k+1\right)!} D_k\left(p,q\right)$$

$$(2.20) \qquad + \sum_{k=0}^{n-1} \frac{(-1)^k}{2^{k+1}\left(k+1\right)!} I_{(\cdot-1)^{k+1}f^{(k+1)}(\cdot)}\left(p,q\right) + \frac{1}{n!}\int_\Gamma \left[q\left(x\right)\right]^{-n+1}$$

$$\times \left(\int_1^{\frac{p(x)}{q(x)}} \left[\frac{1}{2}p\left(x\right) + \left(\frac{1}{2}-t\right)q\left(x\right)\right]^n f^{(n+1)}\left(t\right) dt\right) d\mu\left(x\right).$$

3. Bounds for the Remainder

In this section, we point out some bounds for the remainders in the representations (2.9) and (2.10), i.e.,

$$R_f\left(p,q\right) := (-1)^n \int_\Gamma q\left(x\right)$$

$$(3.1) \qquad \times \left(\int_1^{\frac{p(x)}{q(x)}} K_n\left[\frac{\left(1-\lambda\right)q\left(x\right) + \lambda p\left(x\right)}{q\left(x\right)}, t\right] f^{(n+1)}\left(t\right) dt\right) d\mu\left(x\right)$$

and

$$\tilde{R}_f\left(p,q\right) := \frac{1}{n!}\int_\Gamma \left[q\left(x\right)\right]^{-n+1}$$

$$(3.2) \qquad \times \left(\int_1^{\frac{p(x)}{q(x)}} \left[\lambda p\left(x\right) + \left[\left(1-\lambda\right)-t\right]q\left(x\right)\right]^n f^{(n+1)}\left(t\right) dt\right) d\mu\left(x\right),$$

where $p, q \in \Omega$, $\lambda \in [0,1]$ and $K_n\left(\cdot,\cdot\right)$ is the kernel defined in equation (2.2).

For $a, b \in \mathbb{R}$, let us denote

$$\|f\|_{[a,b],p} := \left|\int_a^b |f\left(t\right)|^p dt\right|^{\frac{1}{p}}, \quad p \geq 1,$$

and

$$\|f\|_{[a,b],\infty} := \mathop{ess\ sup}_{t\in[a,b](t\in[b,a])} |f\left(t\right)|.$$

In order to obtain bounds on $R_f\left(p,q\right)$ as given in (3.1), we need to consider integrals of the form

$$I_1\left(z\right) := \int_1^z K_n\left[\left(1-\lambda\right)\cdot 1 + \lambda z, t\right] f^{(n+1)}\left(t\right) dt, \quad z \in \left(0,\infty\right).$$

Thus we have

$$\left|I_1\left(z\right)\right| \leq \left|\int_1^z \left|K_n\left[(1-\lambda)\cdot 1 + \lambda z, t\right]\right| \left|f^{(n+1)}\left(t\right)\right| dt\right|$$

$$\leq \left\|f^{(n+1)}\right\|_{[1,z],\infty} \left|\int_1^z \left|K_n\left((1-\lambda)\cdot 1 + \lambda z, t\right)\right| dt\right|$$

$$= \frac{1}{n!}\left\|f^{(n+1)}\right\|_{[1,z],\infty} \left|\int_1^{(1-\lambda)\cdot 1+\lambda z} |t-1|^n\, dt + \int_{(1-\lambda)\cdot 1+\lambda z}^z |t-z|^n\, dt\right|$$

$$= \frac{1}{n!}\left[\frac{\left|(1-\lambda)+\lambda z - 1\right|^{n+1} + \left|z - (1-\lambda)\cdot 1 - \lambda z\right|^{n+1}}{n+1}\right]\left\|f^{(n+1)}\right\|_{[1,z],\infty}$$

$$= \frac{1}{n!}\left[\frac{\lambda^{n+1}|z-1|^{n+1} + (1-\lambda)^{n+1}|z-1|^{n+1}}{n+1}\right]\left\|f^{(n+1)}\right\|_{[1,z],\infty}$$

$$= \frac{|z-1|^{n+1}}{(n+1)!}\left[\lambda^{n+1} + (1-\lambda)^{n+1}\right]\left\|f^{(n+1)}\right\|_{[1,z],\infty}.$$

Using Hölder's inequality, we may write, for $\alpha > 1$, $\frac{1}{\alpha} + \frac{1}{\beta} = 1$,

$$\left|I_1\left(z\right)\right| \leq \left\|f^{(n+1)}\right\|_{[1,z],\beta} \left|\int_1^z \left|K_n\left((1-\lambda)\cdot 1 + \lambda z, t\right)\right|^\alpha dt\right|^{\frac{1}{\alpha}}.$$

However,

$$\left|\int_1^z \left|K_n\left((1-\lambda)\cdot 1 + \lambda z, t\right)\right|^\alpha dt\right|^{\frac{1}{\alpha}}$$

$$= \frac{1}{n!}\left|\int_1^{(1-\lambda)\cdot 1+\lambda z} |t-1|^{\alpha n}\, dt + \int_{(1-\lambda)\cdot 1+\lambda z}^z |t-z|^{\alpha n}\, dt\right|^{\frac{1}{\alpha}}$$

$$= \frac{1}{n!}\left[\frac{\left|(1-\lambda)+\lambda z - 1\right|^{\alpha n+1} + \left|z - (1-\lambda)\cdot 1 - \lambda z\right|^{\alpha n+1}}{\alpha n+1}\right]^{\frac{1}{\alpha}}$$

$$= \frac{1}{n!}\left[\frac{\lambda^{\alpha n+1}|z-1|^{\alpha n+1} + (1-\lambda)^{\alpha n+1}|z-1|^{\alpha n+1}}{\alpha n+1}\right]^{\frac{1}{\alpha}}$$

$$= \frac{|z-1|^{n+\frac{1}{\alpha}}}{n!\,(\alpha n+1)^{\frac{1}{\alpha}}}\left[\lambda^{\alpha n+1} + (1-\lambda)^{\alpha n+1}\right]^{\frac{1}{\alpha}}$$

and then we have

$$\left|I_1\left(z\right)\right| \leq \left\|f^{(n+1)}\right\|_{[1,z],\beta}\frac{|z-1|^{n+\frac{1}{\alpha}}}{n!\,(\alpha n+1)^{\frac{1}{\alpha}}}\left[\lambda^{\alpha n+1} + (1-\lambda)^{\alpha n+1}\right]^{\frac{1}{\alpha}}.$$

Finally, we observe that

$$\sup_{t\in[1,z]} |K_n\left((1-\lambda)\cdot 1 + \lambda z, t\right)|$$

$$= \frac{1}{n!}\max\left\{((1-\lambda)+\lambda z - 1)^n + (z - (1-\lambda)\cdot 1 - \lambda z)^n\right\}$$

$$= \frac{1}{n!}(z-1)^n\left(\max\{\lambda, 1-\lambda\}\right)^n$$

$$= \frac{1}{n!}|z-1|^n\left[\frac{1}{2} + \left|\lambda - \frac{1}{2}\right|\right]^n$$

and then

$$|I_1(z)| \le \frac{1}{n!}|z-1|^n\left[\frac{1}{2} + \left|\lambda - \frac{1}{2}\right|\right]^n \left\|f^{(n+1)}\right\|_{[1,z],1}.$$

Using the above inequalities, we may state the following result

$$(3.3) \qquad |I_1(z)| \le \begin{cases} \frac{|z-1|^{n+1}}{(n+1)!}\left[\lambda^{n+1} + (1-\lambda)^{n+1}\right]\left\|f^{(n+1)}\right\|_{[1,z],\infty}, \\[2mm] \frac{|z-1|^{n+\frac{1}{\alpha}}}{n!(\alpha n+1)^{\frac{1}{\alpha}}}\left[\lambda^{\alpha n+1} + (1-\lambda)^{\alpha n+1}\right]^{\frac{1}{\alpha}}\left\|f^{(n+1)}\right\|_{[1,z],\beta} \\[2mm] \qquad \left(\alpha > 1,\ \frac{1}{\alpha} + \frac{1}{\beta} = 1\right), \\[2mm] \frac{1}{n!}|z-1|^n\left[\frac{1}{2} + \left|\lambda - \frac{1}{2}\right|\right]^n\left\|f^{(n+1)}\right\|_{[1,z],1} \end{cases}$$

$$=: \kappa(z,n)$$

for all $z > 0$ and $n \in \mathbb{N}$.

We are now able to state the following theorem referring to the remainder $R_f(p,q)$:

Theorem 2. *Assume that the function f is as in Theorem 1. If $p, q \in \Omega$, then we have the inequality*

$$|R_f(p,q)|$$

$$(3.4) \qquad \le A := \begin{cases} \frac{1}{(n+1)!}\left[\lambda^{n+1} + (1-\lambda)^{n+1}\right]\int_\Gamma\left\{\left[q(x)\right]^{-n} \right. \\[2mm] \qquad \times |p(x) - q(x)|^{n+1}\left.\left\|f^{(n+1)}\right\|_{\left[1,\frac{p(x)}{q(x)}\right],\infty}\right\}d\mu(x), \\[2mm] \frac{1}{n!(\alpha n+1)^{\frac{1}{\alpha}}}\left[\lambda^{\alpha n+1} + (1-\lambda)^{\alpha n+1}\right]^{\frac{1}{\alpha}} \\[2mm] \qquad \times \int_\Gamma\left\{\left[q(x)\right]^{-n-\frac{1}{\alpha}+1}|p(x) - q(x)|^{n+\frac{1}{\alpha}} \right. \\[2mm] \qquad \times\left.\left\|f^{(n+1)}\right\|_{\left[1,\frac{p(x)}{q(x)}\right],\beta}\right\}d\mu(x) \quad \left(\alpha > 1,\ \frac{1}{\alpha} + \frac{1}{\beta} = 1\right), \\[2mm] \frac{1}{n!}\left[\frac{1}{2} + \left|\lambda - \frac{1}{2}\right|\right]^n\int_\Gamma\left[q(x)\right]^{-n+1}|p(x) - q(x)|^n \\[2mm] \qquad \times\left\|f^{(n+1)}\right\|_{\left[1,\frac{p(x)}{q(x)}\right],1}d\mu(x). \end{cases}$$

Moreover, if we assume that $0 \leq r \leq \frac{p(x)}{q(x)} \leq R < \infty$ for $x \in \Gamma$, then the second term in (3.4) can be upper bounded by

$$
(3.5) \quad B := \begin{cases}
\frac{1}{(n+1)!}\left[\lambda^{n+1} + (1-\lambda)^{n+1}\right]\left\|f^{(n+1)}\right\|_{[r,R],\infty} \\
\quad \times \int_\Gamma [q(x)]^{-n}\,|p(x) - q(x)|^{n+1}\,d\mu(x), \\[4pt]
\frac{1}{n!(\alpha n+1)^{\frac{1}{\alpha}}}\left[\lambda^{\alpha n+1} + (1-\lambda)^{\alpha n+1}\right]^{\frac{1}{\alpha}}\left\|f^{(n+1)}\right\|_{[r,R],\beta} \\
\quad \times \int_\Gamma [q(x)]^{-n-\frac{1}{\alpha}+1}\,|p(x) - q(x)|^{n+\frac{1}{\alpha}}\,d\mu(x) \\
\quad (\alpha > 1,\ \frac{1}{\alpha} + \frac{1}{\beta} = 1), \\[4pt]
\frac{1}{n!}\left[\frac{1}{2} + \left|\lambda - \frac{1}{2}\right|\right]^{n}\left\|f^{(n+1)}\right\|_{[r,R],1} \\
\quad \times \int_\Gamma [q(x)]^{-n+1}\,|p(x) - q(x)|^{n}\,d\mu(x)
\end{cases}
$$

$$
\leq C := \begin{cases}
\frac{1}{(n+1)!}\left[\lambda^{n+1} + (1-\lambda)^{n+1}\right]\left\|f^{(n+1)}\right\|_{[r,R],\infty}(R-r)^{n+1}, \\[4pt]
\frac{1}{n!(\alpha n+1)^{\frac{1}{\alpha}}}\left[\lambda^{\alpha n+1} + (1-\lambda)^{\alpha n+1}\right]^{\frac{1}{\alpha}} \\
\quad \times \left\|f^{(n+1)}\right\|_{[r,R],\beta}(R-r)^{n+\frac{1}{\alpha}} \quad (\alpha > 1,\ \frac{1}{\alpha} + \frac{1}{\beta} = 1), \\[4pt]
\frac{1}{n!}\left[\frac{1}{2} + \left|\lambda - \frac{1}{2}\right|\right]^{n}\left\|f^{(n+1)}\right\|_{[r,R],1}(R-r)^{n}.
\end{cases}
$$

The proof of (3.4) follows by the inequality (3.3) choosing $z = \frac{p(x)}{q(x)}$ and integrating. The proof of (3.5) follows by the fact that $\left|\frac{p(x)}{q(x)} - 1\right| \leq R - r$ for all $x \in \Gamma$. We omit the details.

The following corollary may be useful in practical applications.

Corollary 1. *With the assumptions of Theorem 2, we have the inequality:*

$$
(3.6) \quad \left| I_f(p,q) - f(1) - \sum_{k=0}^{n-1} \frac{f^{(k+1)}(1)}{(k+1)!} D_k(p,q) \right|
$$

$$
\leq \begin{cases}
\frac{1}{(n+1)!} \int_\Gamma [q(x)]^{-n}\,|p(x) - q(x)|^{n+1}\left\|f^{(n+1)}\right\|_{\left[1,\frac{p(x)}{q(x)}\right],\infty}\,d\mu(x), \\[4pt]
\frac{1}{n!(\alpha n+1)^{\frac{1}{\alpha}}} \int_\Gamma [q(x)]^{-n-\frac{1}{\alpha}+1}\,|p(x) - q(x)|^{n+\frac{1}{\alpha}} \\
\quad \times \left\|f^{(n+1)}\right\|_{\left[1,\frac{p(x)}{q(x)}\right],\beta}\,d\mu(x), \quad (\alpha > 1,\ \frac{1}{\alpha} + \frac{1}{\beta} = 1), \\[4pt]
\frac{1}{n!} \int_\Gamma [q(x)]^{-n+1}\,|p(x) - q(x)|^{n}\left\|f^{(n+1)}\right\|_{\left[1,\frac{p(x)}{q(x)}\right],1}\,d\mu(x)
\end{cases}
$$

$$
=: M_1.
$$

Furthermore, if we assume that $r \leq \frac{p(x)}{q(x)} \leq R < \infty$, $x \in \Gamma$, then we have

$$M_1 \leq \begin{cases} \frac{1}{(n+1)!} \left\| f^{(n+1)} \right\|_{[r,R],\infty} \int_\Gamma [q(x)]^{-n} |p(x) - q(x)|^{n+1} \, d\mu(x), \\[2mm] \frac{1}{n!(\alpha n+1)^{\frac{1}{\alpha}}} \left\| f^{(n+1)} \right\|_{[r,R],\beta} \int_\Gamma [q(x)]^{-n-\frac{1}{\alpha}+1} \\[2mm] \qquad \times |p(x) - q(x)|^{n+\frac{1}{\alpha}} \, d\mu(x), \quad (\alpha > 1, \frac{1}{\alpha} + \frac{1}{\beta} = 1), \\[2mm] \frac{1}{n!} \left\| f^{(n+1)} \right\|_{[r,R],1} \int_\Gamma [q(x)]^{-n+1} |p(x) - q(x)|^{n} \, d\mu(x) \end{cases}$$

$$=: M_2$$

$$\leq \begin{cases} \frac{1}{(n+1)!} \left\| f^{(n+1)} \right\|_{[r,R],\infty} (R-r)^{n+1}, \\[2mm] \frac{1}{n!(\alpha n+1)^{\frac{1}{\alpha}}} \left\| f^{(n+1)} \right\|_{[r,R],\beta} (R-r)^{n+\frac{1}{\alpha}} \quad (\alpha > 1, \frac{1}{\alpha} + \frac{1}{\beta} = 1), \\[2mm] \frac{1}{n!} \left\| f^{(n+1)} \right\|_{[r,R],1} (R-r)^{n} \end{cases}$$

$$=: M_3$$

and

$$\left| I_f(p,q) - f(1) - \sum_{k=0}^{n-1} \frac{(-1)^k}{(k+1)!} I_{(-1)^{k+1} f^{(k+1)}(\cdot)}(p,q) \right|$$

$$\leq M_1 \leq M_2 \leq M_3$$

and, if $0 \leq r \leq \frac{p(x)}{q(x)} \leq R < \infty$ for $x \in \Gamma$, then

$$\left| I_f(p,q) - f(1) - \sum_{k=0}^{n-1} \left[\frac{1 + (-1)^k}{2^{k+1}(k+1)!} \right] I_{(-1)^{k+1} f^{(k+1)}\left(\frac{1+\cdot}{2}\right)}(p,q) \right|$$

$$\leq \frac{1}{2^n} M_1 \leq \frac{1}{2^n} M_2 \leq \frac{1}{2^n} M_3.$$

Now, to obtain the bound on $\tilde{R}_f(p,q)$ as defined in (3.2), consider the integral

$$I_2(z) := \frac{1}{n!} \int_1^z ((1-\lambda) \cdot 1 + \lambda z - t)^n f^{(n+1)}(t) \, dt.$$

Then we have

$$|I_2(z)|$$

$$\leq \left\| f^{(n+1)} \right\|_{[1,z],\infty} \frac{1}{n!} \left| \int_1^z |(1-\lambda) \cdot 1 + \lambda z - t|^n \, dt \right|$$

$$= \frac{1}{n!} \left| \int_1^{(1-\lambda)\cdot 1 + \lambda z} |(1-\lambda)\cdot 1 + \lambda z - t|^n \, dt \right.$$

$$+ \left. \int_{(1-\lambda)\cdot 1 + \lambda z}^z |(1-\lambda)\cdot 1 + \lambda z - t|^n \, dt \right| \left\| f^{(n+1)} \right\|_{[1,z],\infty}$$

$$= \frac{1}{n!} \cdot \left[\frac{|(1-\lambda)\cdot 1 + \lambda z - 1|^{n+1} + |(1-\lambda)\cdot 1 + \lambda z - z|^{n+1}}{n+1} \right] \left\| f^{(n+1)} \right\|_{[1,z],\infty}$$

$$= \frac{(z-1)^{n+1}}{(n+1)!} \cdot [\lambda^{n+1} + (1-\lambda)^{n+1}] \cdot \left\| f^{(n+1)} \right\|_{[1,z],\infty}.$$

Using Hölder's inequality, we may write, for $\alpha > 1$, $\frac{1}{\alpha} + \frac{1}{\beta} = 1$,

$$|I_2(z)|$$

$$\leq \frac{1}{n!} \left\| f^{(n+1)} \right\|_{[1,z],\beta} \left| \int_1^z |(1-\lambda)\cdot 1 + \lambda z - t|^{n\alpha} \, dt \right|^{\frac{1}{\alpha}}$$

$$= \frac{1}{n!} \left\| f^{(n+1)} \right\|_{[1,z],\beta} \left| \int_1^{(1-\lambda)\cdot 1 + \lambda z} |(1-\lambda)\cdot 1 + \lambda z - t|^{n\alpha} \, dt \right.$$

$$+ \left. \int_{(1-\lambda)\cdot 1 + \lambda z}^z |(1-\lambda)\cdot 1 + \lambda z - t|^{n\alpha} \, dt \right|^{\frac{1}{\alpha}}$$

$$= \frac{1}{n!} \left\| f^{(n+1)} \right\|_{[1,z],\beta} \left[\frac{|(1-\lambda) + \lambda z - 1|^{n\alpha+1} + |z - (1-\lambda) - \lambda z|^{n\alpha+1}}{n\alpha + 1} \right]^{\frac{1}{\alpha}}$$

$$= \frac{1}{n!} \left\| f^{(n+1)} \right\|_{[1,z],\beta} \left[\frac{\lambda^{\alpha n+1} |z-1|^{\alpha n+1} + (1-\lambda)^{\alpha n+1} |z-1|^{\alpha n+1}}{\alpha n + 1} \right]^{\frac{1}{\alpha}}$$

$$= \frac{|z-1|^{n+\frac{1}{\alpha}}}{n! \, (\alpha n + 1)^{\frac{1}{\alpha}}} \cdot [\lambda^{\alpha n+1} + (1-\lambda)^{\alpha n+1}]^{\frac{1}{\alpha}} \cdot \left\| f^{(n+1)} \right\|_{[1,z],\beta}.$$

Finally, we observe that

$$|I_2(z)|$$

$$\leq \frac{1}{n!} \left\| f^{(n+1)} \right\|_{[1,z],1} \sup_{t \in [1,z]} |(1-\lambda)\cdot 1 + \lambda z - t|^n$$

$$= \frac{1}{n!} \left\| f^{(n+1)} \right\|_{[1,z],1} \max \{ |(1-\lambda) + \lambda z - 1|^n + |z - (1-\lambda)\cdot 1 - \lambda z|^n \}$$

$$= \frac{1}{n!} \left\| f^{(n+1)} \right\|_{[1,z],1} (z-1)^n (\max \{\lambda, 1-\lambda\})^n$$

$$= \frac{1}{n!} \left\| f^{(n+1)} \right\|_{[1,z],1} |z-1|^n \left[\frac{1}{2} + \left| \lambda - \frac{1}{2} \right| \right]^n.$$

14   N. S. BARNETT, P. CERONE, S. S. DRAGOMIR ET AL.

Using the above inequalities, we may state that

$$(3.7) \qquad |I_2(z)| \leq \kappa(n, z),$$

where $\kappa(n, z)$ is defined in (3.3). That is, the bounds for $R_f(p, q)$ and $\tilde{R}_f(p, q)$ are the same.

We may now state the following theorem concerning a bound of the remainder $\tilde{R}_f(p, q)$.

Theorem 3. *Assume that the function f is as in Theorem 1. If $p, q \in \Omega$, then we have the inequality:*

$$(3.8) \qquad \left| \tilde{R}_f(p, q) \right| \leq A,$$

where A is given in (3.4). Moreover, if we assume that $0 \leq r \leq \frac{p(x)}{q(x)} \leq R < \infty$, $x \in \Gamma$, then

$$(3.9) \qquad A \leq B \leq C$$

with B and C being as defined in (3.5).

The following corollary may be useful in practical applications:

Corollary 2. *With the above assumptions, we have*

$$(3.10) \qquad \left| I_f(p, q) - f(1) - \sum_{k=0}^{n-1} \frac{(-1)^k}{(k+1)!} I_{(\cdot-1)^{k+1} f^{(k+1)}(\cdot)}(p, q) \right|$$
$$\leq M_1 \leq M_2 \leq M_3,$$

$$(3.11) \qquad \left| I_f(p, q) - f(1) - \sum_{k=0}^{n-1} \frac{f^{(k+1)}(1)}{(k+1)!} D_k(p, q) \right|$$
$$\leq M_1 \leq M_2 \leq M_3$$

and

$$(3.12) \qquad \left| I_f(p, q) - f(1) - \sum_{k=0}^{n-1} \frac{f^{(k+1)}(1)}{2^{k+1}(k+1)!} D_k(p, q) \right.$$
$$\left. - \sum_{k=0}^{n-1} \frac{(-1)^k}{2^{k+1}(k+1)!} I_{(\cdot-1)^{k+1} f^{(k+1)}(\cdot)}(p, q) \right|$$
$$\leq \frac{1}{2^n} M_1 \leq \frac{1}{2^n} M_2 \leq \frac{1}{2^n} M_3$$

for $r \leq \frac{p(x)}{q(x)} \leq R$, $x \in \Gamma$, where M_i $(i = \overline{1,3})$ are as defined in Corollary 1.

Remark 3. If in all the above results we choose f to be a particular function generating the divergences listed at (1.2)~(1.9), then we can obtain many interesting approximations of the above distances. We omit the details.

REFERENCES

1. S. M. Ali and S. D. Silvey, *A general class of coefficients of divergence of one distribution from another*, J. Roy. Statist. Soc. Sec B **28** (1966), 131–142.
2. N. S. Barmett, S. S. Dragomir and A. Sofo, *Better bounds for an inequality of the Ostrowski type with applications*, RGMIA Research Report Collection **3(1)** (2000), Article 1.
3. M. Bethbassat, *f-entropies, probability of error and feature selection*, Inform. Control **39** (1978), 227–242.
4. A. Bhattacharyya, *On a measure of divergence between two statistical populations defined by their probability distributions*, Bull. Calcutta Math. Soc. **35** (1943), 99–109.
5. I. Burbea and C. R. Rao, *On the convexity of some divergence measures based on entropy function*, IEEE Trans. Inf. Th. **28(3)** (1982), 489–495.
6. P. Cerone and S. S. Dragomir, Midpoint type rules from an inequalities point of view, Handbook of Analytic-Computational Methods in Applied Mathematics, Editor: G. Anastassiou, CRC Press, N.Y., 2000, pp. 135–200.
7. P. Cerone and S. S. Dragomir, Trapezoidal type rules from an inequalities point of view, Handbook of Analytic-Computational Methods in Applied Mathematics, Editor: G. Anastassiou, CRC Press, N.Y., 2000, pp. 65–134.
8. P. Cerone, S. S. Dragomir and J. Roumeliotis, *Some Ostrowski type inequalities for n-time differentiable mappings and applications*, Demonstratio Math. **32(2)** (1999), 697–712.
9. P. Cerone, S. S. Dragomir, J. Roumeliotis and J. Šunde, *A new generalization of the trapezoid formula for n-time differentiable mappings and applications*, to appear in Demonstratio Math.
10. C. H. Chen, *Statistical Pattern Recognition*, Rocelle Park, New York, Hoyderc Book Co., 1973.
11. C. K. Chow and C. N. Lin, *Approximating discrete probability distributions with dependence trees*, IEEE Trans. Inf. Th. **14(3)** (1968), 462–467.
12. I. Csiszár, *Information-type measures of difference of probability distributions and indirect observations*, Studia Math. Hungarica **2** (1967), 299–318.
13. I. Csiszár, *A note on Jensen's inequality*, Studia Sci. Math. Hung. **1** (1966), 185–188.
14. I. Csiszár, *On topological properties of f-divergences*, Studia Math. Hungarica **2** (1967), 329–339.
15. I. Csiszár, *A note on Jensen's inequality*, Studia Sci. Math. Hung. **1** (1966), 185-188..
16. I. Csiszár and J. Körner, *Information Theory: Coding Theorem for Discrete Memoryless Systems*, Academic Press, New York, 1981.
17. D. Dacunha-Castelle, *Ecole d' ete de Probability de Saint-Flour, III-1977*, Springer Verlag, Berlin, Heidelberg, 1978.
18. S. S. Dragomir and S. Mabizela, *Some error estimates in the trapezoidal quadrature rule*, RGMIA Research Report Collect. **2(5)** (1999), Article 6 (http://rgmia.vu.edu.au/v2n5.html).
19. S. S. Dragomir and S. Wang, *A new inequality of Ostrowski-Grüss type and its applications to the estimation of error bounds for special means and for some numerical quadrature rules*, Comp. Math. Appl. **33(11)** (1997), 15–20.
20. D. V. Gokhale and S. Kullback, *Information in Contingency Tables*, New York, Marcel Dekker, 1978.

21. J. H. Havrda and F. Charvat, *Quantification method classification process: concept of structural α-entropy*, Kybernetika **3** (1967), 30–35.

22. E. Hellinger, *Neue Bergrüirdung du Theorie quadratisher Formerus von uneudlichvieleu Veränderlicher*, J. für reine and Augeur. Math. **36** (1909), 210–271.

23. H. Jeffreys, *An invariant form for the prior probability in estimating problems*, Proc. Roy. Soc. London **186A** (1946), 453–461.

24. T. T. Kadota and L.A. Shepp, *On the best finite set of linear observables for discriminating two Gaussian signals*, IEEE Trans. Inf. Th. **13** (1967), 288–294.

25. T. Kailath, *The divergence and Bhattacharyya distance measures in signal selection*, IEEE Trans. Comm. Technology. **COM-15** (1967), 52–60.

26. J. N. Kapur, *A comparative assessment of various measures of directed divergence*, Advances in Management Studies **3** (1984), 1–16.

27. D. Kazakos and T. Cotsidas, *A decision theory approach to the approximation of discrete probability densities*, IEEE Trans. Perform. Anal. Machine Intell. **1** (1980), 61–67.

28. J. H. B. Kemperman, *On the optimum note of transmitting information*, Ann. Math. Statist. **40** (1969), 2158–2177.

29. C. Kraft, *Some conditions for consistency and uniform consistency of statistical procedures*, Univ. of California Pub. in Statistics **1** (1955), 125–142.

30. S. Kullback, *A lower bound for discrimination information in terms of variation*, IEEE Trans. Inf. Th. **13** (1967), 126–127.

31. S. Kullback, *Correction to a lower bound for discrimination information in terms of variation*, IEEE Trans. Inf. Th. **16** (1970), 771–773.

32. S. Kullback and R. A. Leibler, *On information and sufficiency*, Ann. Math. Stat. **22** (1951), 79–86.

33. L. Lecam, *Asymptotic Methods in Statistical Decision Theory*, Springer Verlag, New York, 1986.

34. J. Lin, *Divergence measures based on the Shannon entropy*, IEEE Trans. Inf. Th. **37(1)** (1991), 145–151.

35. J. Lin and S. K. M. Wong, *A new directed divergence measure and its characterization*, Int. J. General Systems **17** (1990), 73–81.

36. H. P. Mckean, Jr., *Speed of approach to equilibrium for Koc's caricature of a Maximilian gas*, Arch. Ration. Mech. Anal. **21** (1966), 343-367.

37. M. Mei, *The theory of genetic distance and evaluation of human races*, Japan J. Human Genetics **23** (1978), 341–369.

38. C. E. M. Pearce, J. Pečarić, N. Ujević and S. Varošanec, *Generaliasations of some inequalities of Ostrowski-Grüss type*, Math. Ineq. & Appl. **3(1)** (2000), 25–34.

39. E. C. Pielou, *Ecological Diversity*, Wiley, New York, 1975.

40. M. S. Pinsker, *Information and Information Stability of Random variables and processes*, *(in Russian)*, Moscow: Izv. Akad. Nouk, 1960.

41. C. R. Rao, *Diversity and dissimilarity coefficients: a unified approach*, Theoretic Population Biology **21** (1982), 24–43.

42. A. Rényi, *On measures of entropy and information*, Proc. Fourth Berkeley Symp. Math. Stat. and Prob., University of California Press **1** (1961), 547–561.

43. A. Sen, *On Economic Inequality*, Oxford University Press, London, 1973.

44. B. D. Sharma and D. P. Mittal, *New non-additive measures of relative information*, Journ. Comb. Inf. Sys. Sci. **2(4)** (1977), 122–132.

45. H. ShioyA and T. Da-te, *A generalisation of Lin divergence and the derivative of a new information divergence*, Elec. and Comm. in Japan **787** (1995), 37–40.

46. I. J. Taneja, *Generalised Information Measures and their Applications*, (http://www.mtm.ufsc.brtaneja/bhtml/bhtml.html).
47. H. Theil, *Economics and Information Theory*, North-Holland, Amsterdam, 1967.
48. H. Theil, *Statistical Decomposition Analysis*, North-Holland, Amsterdam, 1972.
49. F. Topsoe, *Some inequalities for information divergence and related measures of discrimination*, Res. Rep. Coll., RGMIA **2** **(1)** (1999), 85–98.
50. G. T. Toussaint, *Sharper lower bounds for discrimination in terms of variation*, IEEE Trans. Inf. Th. **21** (1975), 99–100.
51. I. Vajda, *Note on discrimination information and variation*, IEEE Trans. Inf. Th. **16** (1970), 771–773.
52. I. Vajda, *Theory of Statistical Inference and Information*, Dordrecht-Boston, Kluwer Academic Publishers, 1989.
53. V. A. Volkonski and J. A. Rozanov, *Some limit theorems for random function I, (English Trans.)*, Theory Prob. Appl., (USSR) **4** (1959), 178–197.

APPROXIMATING CSISZÁR f-DIVERGENCE VIA AN OSTROWSKI TYPE IDENTITY FOR n-TIME DIFFERENTIABLE FUNCTIONS

N. S. Barnett[1], P. Cerone[1], S. S. Dragomir[1] and A. Sofo[1]

Abstract. Using an identity of the Ostrowski type for n-differentiable functions, we point out an approximation of Csiszár f-divergence.

1. Introduction

One of the important issues in many applications of Probability Theory is finding an appropriate measure of *distance* (or *difference* or *discrimination*) between two probability distributions. A number of divergence measures for this purpose have been proposed and extensively studied by Jeffreys [25], Kullback and Leibler [34], Rényi [45], Havrda and Charvat [23], Kapur [28], Sharma and Mittal [46], Burbea and Rao [5], Rao [44], Lin [37], Csiszár [11], Ali and Silvey [1], Vajda [55], Shioya and Da-te [48] and others (see for example [28] and the references therein).

These measures have been applied in a variety of fields such as: anthropology [44], genetics [40], finance, economics, and political science [46]. [50]. [51], biology [42], the analysis of contingency tables [22], approximation of probability distributions [10], [29], signal processing [26], [27] and pattern recognition [3], [9].

Assume that a set χ and the σ-finite measure μ are given. Consider the set of all probability densities on μ to be $\Omega := \left\{ p \mid p : \chi \to \mathbb{R}, p(x) \geq \right.$

Received March 16, 2001.

1991 *Mathematics Subject Classification.* Primary 94A99; Secondary 26D15.

Key Words and Phrases. Csiszár f-divergence, analytic inequalities.

[1] School of Communications and Informatics, Victoria University of Technology, PO Box 14428, Melbourne, MC 8001, Australia. Email addresses: neil@matilda.vu.edu.au http://sci.vu.edu.au/staff/neilb.html, pc@matilda.vu.edu.au http://sci.vu.edu.au/staff/peterc.html, sever@matilda.vu.edu.au http:// rgmia.vu.edu.au/SSDragomirWeb.html and sofo@matilda.vu.edu.au http://cams.vu.edu.au/staff/anthonys.html

$0,\ \int_{\chi} p(x)\, d\mu(x) = 1\}$. The Kullback-Leibler divergence [34] is well known among the information divergences. It is defined as:

$$(1.1) \qquad D_{KL}(p,q) := \int_{\chi} p(x) \log\left[\frac{p(x)}{q(x)}\right] d\mu(x), \quad p,q \in \Omega,$$

where log is to base 2.

In Information Theory and Statistics, various divergences are applied in addition to the Kullback-Leibler divergence. These are the: *variation distance* D_v, *Hellinger distance* D_H [24], χ^2-*divergence* D_{χ^2}, α-*divergence* D_α, *Bhattacharyya distance* D_B [4], *Harmonic distance* D_{Ha}, *Jeffreys distance* D_J [25], *triangular discrimination* D_Δ [52], etc. They are defined as follows, respectively:

$$(1.2) \qquad D_v(p,q) := \int_{\chi} |p(x) - q(x)|\, d\mu(x), \quad p,q \in \Omega,$$

$$(1.3) \qquad D_H(p,q) := \int_{\chi} \left[\sqrt{p(x)} - \sqrt{q(x)}\right]^2 d\mu(x), \quad p,q \in \Omega,$$

$$(1.4) \qquad D_{\chi^2}(p,q) := \int_{\chi} p(x)\left[\left(\frac{q(x)}{p(x)}\right)^2 - 1\right] d\mu(x), \quad p,q \in \Omega,$$

$$(1.5) \quad D_\alpha(p,q) := \frac{4}{1-\alpha^2}\left[1 - \int_{\chi} [p(x)]^{\frac{1-\alpha}{2}} [q(x)]^{\frac{1+\alpha}{2}}\, d\mu(x)\right], \quad p,q \in \Omega,$$

$$(1.6) \qquad D_B(p,q) := \int_{\chi} \sqrt{p(x)\, q(x)}\, d\mu(x), \quad p,q \in \Omega,$$

$$(1.7) \qquad D_{Ha}(p,q) := \int_{\chi} \frac{2p(x)\, q(x)}{p(x) + q(x)}\, d\mu(x), \quad p,q \in \Omega,$$

$$(1.8) \qquad D_J(p,q) := \int_{\chi} [p(x) - q(x)] \ln\left[\frac{p(x)}{q(x)}\right] d\mu(x), \quad p,q \in \Omega,$$

$$(1.9) \qquad D_\Delta(p,q) := \int_{\chi} \frac{[p(x) - q(x)]^2}{p(x) + q(x)}\, d\mu(x), \quad p,q \in \Omega.$$

For other divergence measures, see the paper [28] by Kapur or the book on line [49] by Taneja. For a comprehensive collection of preprints available on line, see the RGMIA web site *http://rgmia.vu.edu.au/papersinfth.html*

Csiszár f-divergence is defined as follows [11]:

$$(1.10) \qquad I_f(p,q) := \int_{\chi} q(x) f\left[\frac{p(x)}{q(x)}\right] d\mu(x), \quad p,q \in \Omega,$$

where f is convex on $(0, \infty)$. It is assumed that $f(u)$ is zero and strictly convex at $u = 1$. By appropriately defining this convex function, various divergences are derived. All the above distances $(1.1)\sim(1.9)$, are particular instances of Csiszár f-divergence. There are also many others which are not in this class (see for example [28] or [49]).

2. Representation of Csiszár f-Divergence

In [8], the authors have pointed out the following integral identity generalising the trapezoid rule.

Lemma 1. *Let* $f : [a, b] \to \mathbb{R}$ *be a mapping such that* $f^{(n-1)}$ *is absolutely continuous on* $[a, b]$. *Then for all* $x \in [a, b]$, *we have the identity:*

$$
(2.1) \quad \int_a^b f(t)\, dt = \sum_{k=0}^{n-1} \left[\frac{(b-x)^{k+1} + (-1)^k (x-a)^{k+1}}{(k+1)!} \right] f^{(k)}(x)
$$

$$
+ (-1)^n \int_a^b K_n(x, t)\, f^{(n)}(t)\, dt,
$$

where the kernel $K_n : [a, b]^2 \to \mathbb{R}$ *is given by*

$$
(2.2) \quad K_n(x, t) := \begin{cases} \frac{(t-a)^n}{n!}, & a \le t \le x \le b, \\ \frac{(t-b)^n}{n!}, & a \le x < t \le b \end{cases}
$$

and n *is a natural number,* $n \ge 1$.

In what follows, we need the identity (2.2) in the following equivalent form:

$$
(2.3) \quad \begin{aligned} f(z) &= \frac{1}{b-a} \int_a^b f(t)\, dt \\ &\quad - \frac{1}{b-a} \sum_{k=1}^{n-1} \frac{1}{(k+1)!} \left[(b-z)^{k+1} + (-1)^k (z-a)^{k+1} \right] f^{(k)}(z) \\ &\quad + \frac{(-1)^{n+1}}{(b-a)\, n!} \left[\int_a^z (t-a)^n f^{(n)}(t)\, dt + \int_z^b (t-b)^n f^{(n)}(t)\, dt \right] \end{aligned}
$$

for all $z \in [a, b]$.

Note that, for $n = 1$, the sum $\sum_{k=1}^{n-1}$ is empty and we obtain the known identity (see for example [6])

$$
(2.4) \quad \begin{aligned} f(z) &= \frac{1}{b-a} \int_a^b f(t)\, dt + \frac{1}{b-a} \left[\int_a^z (t-a) f^{(1)}(t)\, dt \right. \\ &\quad \left. + \int_z^b (t-b) f^{(1)}(t)\, dt \right], \quad x \in [a, b]. \end{aligned}
$$

In what follows, we assume that the probability distributions $p, q \in \Omega$ satisfy the standing condition:

$$(2.5) \qquad 0 \leq r \leq \frac{p(x)}{q(x)} \leq R < \infty, \quad a.e. \ x \in \Gamma.$$

Obviously, $r \leq 1 \leq R$.

The following representation of Csiszár f–divergence holds:

Theorem 1. *Let* $f : [r, R] \to \mathbb{R}$, *where* r, R *are as above and* $f^{(n-1)}$ *is absolutely continuous on* $[r, R]$. *Then for all* $p, q \in \Omega$ *satisfying* (2.5), *we have the representation:*

$$
\begin{aligned}
I_f\left(p, q\right) \\
= \frac{1}{R-r} \int_r^R f\left(t\right) dt &- \frac{1}{R-r} \sum_{k=1}^{n-1} \frac{1}{(k+1)!} I_{(R-\cdot)^{k+1} f^{(k)}(\cdot)}\left(p, q\right) \\
+ \frac{1}{R-r} \sum_{k=1}^{n-1} \frac{(-1)^{k+1}}{(k+1)!} & I_{(\cdot-r)^{k+1} f^{(k)}(\cdot)}\left(p, q\right) \\
+ \frac{(-1)^{n+1}}{(R-r)\, n!} \int_\Gamma q\left(x\right) & \left(\int_r^{\frac{p(x)}{q(x)}} \left(t-r\right)^n f^{(n)}\left(t\right) dt \right) d\mu\left(x\right) \\
+ \frac{(-1)^{n+1}}{(R-r)\, n!} \int_\Gamma q\left(x\right) & \left(\int_{\frac{p(x)}{q(x)}}^R \left(t-R\right)^n f^{(n)}\left(t\right) dt \right) d\mu\left(x\right).
\end{aligned}
$$
(2.6)

Proof. Using (2.3) for $z = \frac{p(x)}{q(x)}$, $x \in \Gamma$ and $a = r$, $b = R$, we may write

$$
\begin{aligned}
f\left(\frac{p(x)}{q(x)}\right) \\
= \frac{1}{R-r} \int_r^R f\left(t\right) dt &- \frac{1}{R-r} \sum_{k=1}^{n-1} \frac{1}{(k+1)!} \left[\left(R - \frac{p(x)}{q(x)} \right)^{k+1} \right. \\
+ (-1)^k \left(\frac{p(x)}{q(x)} - r \right)^{k+1} & \left. \right] f^{(k)}\left(\frac{p(x)}{q(x)}\right) \\
+ \frac{(-1)^{n+1}}{(R-r)\, n!} \left[\int_r^{\frac{p(x)}{q(x)}} \left(t-r\right)^n f^{(n)}\left(t\right) dt \right. &+ \left. \int_{\frac{p(x)}{q(x)}}^R \left(t-R\right)^n f^{(n)}\left(t\right) dt \right]
\end{aligned}
$$
(2.7)

for all $x \in \Gamma$. If we multiply (2.7) by $q\left(x\right) \geq 0$, integrate on Γ and take into account that $\int_\Gamma q\left(x\right) d\mu\left(x\right) = 1$, we get the desired identity (2.6). This completes the proof.

Remark 1. If $n = 1$, then we have the representation:

$$
\begin{aligned}
I_f(p,q) &= \frac{1}{R-r} \int_r^R f(t)\,dt \\
&\quad + \frac{1}{R-r} \int_\Gamma q(x) \left(\int_r^{\frac{p(x)}{q(x)}} (t-r) f'(t)\,dt \right) d\mu(x) \\
&\quad + \frac{1}{R-r} \int_\Gamma q(x) \left(\int_{\frac{p(x)}{q(x)}}^R (t-R) f'(t)\,dt \right) d\mu(x).
\end{aligned}
$$
(2.8)

If $n = 2$, then we have the representation:

$$
\begin{aligned}
I_f(p,q) &= \frac{1}{R-r} \int_r^R f(t)\,dt + \frac{1}{2(R-r)} I_{(R-\cdot)^2 f^{(1)}(\cdot)}(p,q) \\
&\quad + \frac{1}{R-r} I_{(\cdot-r)^2 f'(\cdot)}(p,q) \\
&\quad - \frac{1}{2(R-r)} \int_\Gamma q(x) \left(\int_r^{\frac{p(x)}{q(x)}} (t-r)^2 f^{(2)}(t)\,dt \right) d\mu(x) \\
&\quad - \frac{1}{2(R-r)} \int_\Gamma q(x) \left(\int_{\frac{p(x)}{q(x)}}^R (t-R)^2 f^{(2)}(t)\,dt \right) d\mu(x).
\end{aligned}
$$
(2.9)

3. Bounds for the Remainder

In the formula (2.6), we consider the remainder $R_f(p,q)$ given by

$$
\begin{aligned}
R_f(p,q) &:= \frac{(-1)^{n+1}}{(R-r)\,n!} \left[\int_\Gamma q(x) \left(\int_r^{\frac{p(x)}{q(x)}} (t-r)^n f^{(n)}(t)\,dt \right) d\mu(x) \right. \\
&\quad \left. + \int_\Gamma q(x) \left(\int_{\frac{p(x)}{q(x)}}^R (t-R)^n f^{(n)}(t)\,dt \right) d\mu(x) \right].
\end{aligned}
$$

In this section, we are interested in obtaining some bounds for $R_f(p,q)$. For this purpose, consider

$$
I_1(z) = \int_r^z (t-r)^n f^{(n)}(t)\,dt
$$

and

$$
I_2(z) := \int_z^R (t-R)^n f^{(n)}(t)\,dt,
$$

where $z \in [r, R]$. For $a < b$, we also define the *Lebesgue norms* as follows,

$$\|f\|_{[a,b],p} := \left[\int_a^b |f(t)|^p \, dt \right]^{\frac{1}{p}}, \quad p \geq 1,$$

and

$$\|f\|_{[a,b],\infty} := ess \sup_{t \in [a,b]} |f(t)|.$$

Now, we observe that

$$|I_1(z)| \leq \int_r^z |t - r|^n \left| f^{(n)}(t) \right| dt \leq \|f^{(n)}\|_{[r,z],\infty} \frac{(z - r)^{n+1}}{n + 1},$$

$$|I_1(z)| \leq \|f^{(n)}\|_{[r,z],\beta} \left(\int_r^z |t - r|^{\alpha n} \, dt \right)^{\frac{1}{\alpha}}$$

$$= \|f^{(n)}\|_{[r,z],\beta} \left[\frac{(z - r)^{\alpha n + 1}}{\alpha n + 1} \right]^{\frac{1}{\alpha}}$$

$$= \|f^{(n)}\|_{[r,z],\beta} \frac{(z - r)^{n + \frac{1}{\alpha}}}{(\alpha n + 1)^{\frac{1}{\alpha}}}, \quad \left(\alpha > 1, \; \frac{1}{\alpha} + \frac{1}{\beta} = 1 \right),$$

and, finally,

$$|I_1(z)| \leq \|f^{(n)}\|_{[r,z],1} \sup_{t \in [r,z]} |t - r|^n = \|f^{(n)}\|_{[r,z],1} (z - r)^n.$$

Consequently, we have

$$(3.1) \qquad |I_1(z)| \leq \begin{cases} \frac{(z-r)^{n+1}}{n+1} \|f^{(n)}\|_{[r,z],\infty}, \\[2mm] \frac{(z-r)^{n+\frac{1}{\alpha}}}{(\alpha n+1)^{\frac{1}{\alpha}}} \|f^{(n)}\|_{[r,z],\beta}, \quad (\alpha > 1, \; \frac{1}{\alpha} + \frac{1}{\beta} = 1), \\[2mm] (z - r)^n \|f^{(n)}\|_{[r,z],1}. \end{cases}$$

In a similar fashion, we may point out that

$$(3.2) \qquad |I_2(z)| \leq \begin{cases} \frac{(R-z)^{n+1}}{n+1} \|f^{(n)}\|_{[z,R],\infty}, \\[2mm] \frac{(R-z)^{n+\frac{1}{\alpha}}}{(\alpha n+1)^{\frac{1}{\alpha}}} \|f^{(n)}\|_{[z,R],\beta}, \quad (\alpha > 1, \; \frac{1}{\alpha} + \frac{1}{\beta} = 1), \\[2mm] (R - z)^n \|f^{(n)}\|_{[z,R],1}. \end{cases}$$

The following bound for the remainder $R_f(p, q)$ holds:

Theorem 2. *Let f, r, R, p and q be as in Theorem 1. Then we have the inequality*

$$|R_f(p,q)|$$

$$\leq \frac{1}{n!(R-r)} \times \begin{cases} \frac{1}{n+1}\int_\Gamma q(x)\left[\left(R-\frac{p(x)}{q(x)}\right)^{n+1}\left\|f^{(n)}\right\|_{\left[\frac{p(x)}{q(x)},R\right],\infty}\right. \\[2mm] \left. +\left(\frac{p(x)}{q(x)}-r\right)^{n+1}\left\|f^{(n)}\right\|_{\left[r,\frac{p(x)}{q(x)}\right],\infty}\right]d\mu(x), \\[3mm] \frac{1}{(\alpha n+1)^{\frac{1}{\alpha}}}\int_\Gamma q(x)\left[\left(R-\frac{p(x)}{q(x)}\right)^{n+\frac{1}{\alpha}}\left\|f^{(n)}\right\|_{\left[\frac{p(x)}{q(x)},R\right],\beta}\right. \\[2mm] \left. +\left(\frac{p(x)}{q(x)}-r\right)^{n+\frac{1}{\alpha}}\left\|f^{(n)}\right\|_{\left[r,\frac{p(x)}{q(x)}\right],\beta}\right]d\mu(x) \\[2mm] \left(\alpha>1,\ \frac{1}{\alpha}+\frac{1}{\beta}=1\right), \\[3mm] \int_\Gamma q(x)\left[\left(R-\frac{p(x)}{q(x)}\right)^{n}\left\|f^{(n)}\right\|_{\left[\frac{p(x)}{q(x)},R\right],1}\right. \\[2mm] \left. +\left(\frac{p(x)}{q(x)}-r\right)^{n}\left\|f^{(n)}\right\|_{\left[r,\frac{p(x)}{q(x)}\right],1}\right]d\mu(x) \end{cases}$$

$$(3.3)$$

$$\leq \frac{1}{n!(R-r)} \times \begin{cases} \frac{\left\|f^{(n)}\right\|_{[r,R],\infty}}{n+1}\int_\Gamma q(x)\left[\left(R-\frac{p(x)}{q(x)}\right)^{n+1}\right. \\[2mm] \left. +\left(\frac{p(x)}{q(x)}-r\right)^{n+1}\right]d\mu(x), \\[3mm] \frac{\left\|f^{(n)}\right\|_{[r,R],\beta}}{(\alpha n+1)^{\frac{1}{\alpha}}}\int_\Gamma q(x)\left[\left(R-\frac{p(x)}{q(x)}\right)^{n+\frac{1}{\alpha}}\right. \\[2mm] \left. +\left(\frac{p(x)}{q(x)}-r\right)^{n+\frac{1}{\alpha}}\right]d\mu(x), \\[3mm] \left\|f^{(n)}\right\|_{[r,R],1}\int_\Gamma q(x)\left[\left(R-\frac{p(x)}{q(x)}\right)^{n}\right. \\[2mm] \left. +\left(\frac{p(x)}{q(x)}-r\right)^{n}\right]d\mu(x) \end{cases}$$

$$\leq \begin{cases} \frac{\left\|f^{(n)}\right\|_{[r,R],\infty}(R-r)^{n}}{(n+1)!}, \\[3mm] \frac{\left\|f^{(n)}\right\|_{[r,R],\beta}(R-r)^{n+\frac{1}{\alpha}-1}}{n!(\alpha n+1)^{\frac{1}{\alpha}}} \quad \left(\alpha>1,\ \frac{1}{\alpha}+\frac{1}{\beta}=1\right), \\[3mm] \frac{\left\|f^{(n)}\right\|_{[r,R],1}(R-r)^{n-1}}{n!}. \end{cases}$$

Proof. Using (3.1) and (3.2), we may write:

$$\|R_f(p,q)\| \leq \frac{1}{n!(R-r)}\left[\int_\Gamma q(x)\left|\int_r^{\frac{p(x)}{q(x)}}(t-r)^n f^{(n)}(t)dt\right|d\mu(x)\right.$$

$$\left. +\int_\Gamma q(x)\left|\int_{\frac{p(x)}{q(x)}}^{R}(t-R)^n f^{(n)}(t)dt\right|d\mu(x)\right]$$

$$\leq \frac{1}{(R-r)\,n!} \times \begin{cases} \frac{1}{n+1} \int_\Gamma q\left(x\right)\left(\frac{p(x)}{q(x)}-r\right)^{n+1}\left\|f^{(n)}\right\|_{\left[r,\frac{p(x)}{q(x)}\right],\infty} d\mu\left(x\right), \\[2mm] \frac{1}{(\alpha n+1)^{\frac{1}{\alpha}}} \int_\Gamma q(x)\left(\frac{p(x)}{q(x)}-r\right)^{n+\frac{1}{\alpha}}\left\|f^{(n)}\right\|_{\left[r,\frac{p(x)}{q(x)}\right],\beta} d\mu\left(x\right), \\[2mm] \qquad \left(\alpha>1,\ \frac{1}{\alpha}+\frac{1}{\beta}=1\right), \\[2mm] \int_\Gamma q\left(x\right)\left(\frac{p(x)}{q(x)}-r\right)^{n}\left\|f^{(n)}\right\|_{\left[r,\frac{p(x)}{q(x)}\right],1} d\mu\left(x\right), \end{cases}$$

$$+ \frac{1}{(R-r)\,n!} \times \begin{cases} \frac{1}{n+1} \int_\Gamma q\left(x\right)\left(R-\frac{p(x)}{q(x)}\right)^{n+1}\left\|f^{(n)}\right\|_{\left[\frac{p(x)}{q(x)},R\right],\infty} d\mu\left(x\right), \\[2mm] \frac{1}{(\alpha n+1)^{\frac{1}{\alpha}}} \int_\Gamma q\left(x\right)\left(R-\frac{p(x)}{q(x)}\right)^{n+\frac{1}{\alpha}}\left\|f^{(n)}\right\|_{\left[\frac{p(x)}{q(x)},R\right],\beta} d\mu\left(x\right), \\[2mm] \qquad \left(\alpha>1,\ \frac{1}{\alpha}+\frac{1}{\beta}=1\right), \\[2mm] \int_\Gamma q\left(x\right)\left(R-\frac{p(x)}{q(x)}\right)^{n}\left\|f^{(n)}\right\|_{\left[\frac{p(x)}{q(x)},R\right],1} d\mu\left(x\right), \end{cases}$$

$$\leq \frac{1}{(R-r)\,n!} \times \begin{cases} \frac{\left\|f^{(n)}\right\|_{[r,R],\infty}}{n+1} \int_\Gamma q\left(x\right)\left[\left(\frac{p(x)}{q(x)}-r\right)^{n+1}\right. \\[2mm] \left.\quad +\left(R-\frac{p(x)}{q(x)}\right)^{n+1}\right]d\mu\left(x\right), \\[2mm] \frac{\left\|f^{(n)}\right\|_{[r,R],\beta}}{(\alpha n+1)^{\frac{1}{\alpha}}} \int_\Gamma q\left(x\right)\left[\left(\frac{p(x)}{q(x)}-r\right)^{n+\frac{1}{\alpha}}\right. \\[2mm] \left.\quad +\left(R-\frac{p(x)}{q(x)}\right)^{n+\frac{1}{\alpha}}\right]d\mu\left(x\right), \\[2mm] \left\|f^{(n)}\right\|_{[r,R],1} \int_\Gamma q\left(x\right)\left[\left(\frac{p(x)}{q(x)}-r\right)^{n}\right. \\[2mm] \left.\quad +\left(R-\frac{p(x)}{q(x)}\right)^{n}\right]d\mu\left(x\right) \end{cases}$$

$$\leq \frac{1}{(R-r)\,n!} \times \begin{cases} \frac{\left\|f^{(n)}\right\|_{[r,R],\infty}}{n+1}\left(R-r\right)^{n+1}, \\[2mm] \frac{\left\|f^{(n)}\right\|_{[r,R],\beta}(R-r)^{n+\frac{1}{\alpha}}}{(\alpha n+1)^{\frac{1}{\alpha}}}, \\[2mm] \left\|f^{(n)}\right\|_{[r,R],1}\left(R-r\right)^{n} \end{cases}$$

$$= \begin{cases} \frac{\left\|f^{(n)}\right\|_{[r,R],\infty}(R-r)^{n}}{(n+1)!}, \\[2mm] \frac{\left\|f^{(n)}\right\|_{[r,R],\beta}(R-r)^{n+\frac{1}{\alpha}-1}}{n!(\alpha n+1)^{\frac{1}{\alpha}}}, \\[2mm] \frac{\left\|f^{(n)}\right\|_{[r,R],1}(R-r)^{n-1}}{n!}. \end{cases}$$

and the theorem is proved. This completes the proof.

Remark 2. For $n = 1$, we obtain the estimate

$$|R_f(p,q)| \leq \frac{1}{R-r} \times \begin{cases} \frac{1}{2} \int_\Gamma q(x) \left[\left(R - \frac{p(x)}{q(x)}\right)^2 \|f^{(1)}\|_{\left[\frac{p(x)}{q(x)}, R\right], \infty} \right. \\ \left. + \left(\frac{p(x)}{q(x)} - r\right)^2 \|f^{(1)}\|_{\left[r, \frac{p(x)}{q(x)}\right], \infty} \right] d\mu(x) \\[2mm] \frac{1}{(\alpha+1)^{\frac{1}{\alpha}}} \int_\Gamma q(x) \left[\left(R - \frac{p(x)}{q(x)}\right)^{1+\frac{1}{\alpha}} \|f^{(1)}\|_{\left[\frac{p(x)}{q(x)}, R\right], \beta} \right. \\ \left. + \left(\frac{p(x)}{q(x)} - r\right)^{1+\frac{1}{\alpha}} \|f^{(1)}\|_{\left[r, \frac{p(x)}{q(x)}\right], \beta} \right] d\mu(x) \\ \quad (\alpha > 1, \frac{1}{\alpha} + \frac{1}{\beta} = 1), \\[2mm] \int_\Gamma q(x) \left[\left(R - \frac{p(x)}{q(x)}\right) \|f^{(1)}\|_{\left[\frac{p(x)}{q(x)}, R\right], 1} \right. \\ \left. + \left(\frac{p(x)}{q(x)} - r\right) \|f^{(1)}\|_{\left[r, \frac{p(x)}{q(x)}\right], 1} \right] d\mu(x) \end{cases}$$

$$\leq \frac{1}{R-r} \times \begin{cases} \frac{\|f^{(1)}\|_{[r,R], \infty}}{2} \int_\Gamma q(x) \left[\left(R - \frac{p(x)}{q(x)}\right)^2 \right. \\ \left. + \left(\frac{p(x)}{q(x)} - r\right)^2 \right] d\mu(x), \\[2mm] \frac{\|f^{(1)}\|_{[r,R], \beta}}{(\alpha+1)^{\frac{1}{\alpha}}} \int_\Gamma q(x) \left[\left(\frac{p(x)}{q(x)} - r\right)^{1+\frac{1}{\alpha}} \right. \\ \left. + \left(R - \frac{p(x)}{q(x)}\right)^{1+\frac{1}{\alpha}} \right] d\mu(x), \quad (\alpha > 1, \frac{1}{\alpha} + \frac{1}{\beta} = 1), \\[2mm] (R - r) \|f^{(1)}\|_{[r,R], 1} \end{cases}$$

$$\leq \begin{cases} \frac{\|f^{(1)}\|_{[r,R], \infty}(R-r)}{2}, \\[2mm] \frac{\|f^{(1)}\|_{[r,R], \beta}(R-r)^{\frac{1}{\alpha}}}{(\alpha+1)^{\frac{1}{\alpha}}} \quad (\alpha > 1, \frac{1}{\alpha} + \frac{1}{\beta} = 1), \\[2mm] \|f^{(1)}\|_{[r,R], 1}, \end{cases}$$

which improves some results from [18] and [19].

REFERENCES

1. S. M. Ali and S. D. Silvey, *A general class of coefficients of divergence of one distribution from another*, J. Roy. Statist. Soc. Sec B **28** (1966), 131–142.
2. N. S. Barnett, S. S. Dragomir and A. Sofo, *Better bounds for an inequality of the Ostrowski type with applications*, Demonstratio Mathematica **34(3)** (2001), 533–542.
3. M. Bethbassat, *f-entropies, probability of error and feature selection*, Inform. Control **39** (1978), 227–242.
4. A. Bhattacharyya, *On a measure of divergence between two statistical populations defined by their probability distributions*, Bull. Calcutta Math. Soc. **35** (1943), 99–109.

5. I. Burbea and C. R. Rao, *On the convexity of some divergence measures based on entropy function*, IEEE Trans. Inf. Th. **28(3)** (1982), 489–495.

6. P. Cerone and S. S. Dragomir, Midpoint type rules from an inequalities point of view, Handbook of Analytic-Computational Methods in Applied Mathematics, Editor: G. Anastassiou, CRC Press, N.Y., 2000, pp. 135–200.

7. P. Cerone and S. S. Dragomir, Trapezoidal type rules from an inequalities point of view, Handbook of Analytic-Computational Methods in Applied Mathematics, Editor: G. Anastassiou, CRC Press, N.Y., 2000, pp. 65–134.

8. P. Cerone, S. S. Dragomir and J. Roumeliotis, *Some Ostrowski type inequalities for n-time differentiable mappings and applications*, Demonstratio Math. **32(2)** (1999), 697–712.

9. C. H. Chen, *Statistical Pattern Recognition*, Rocelle Park, New York, Hoyderc Book Co., 1973.

10. C. K. Chow and C. N. Lin, *Approximating discrete probability distributions with dependence trees*, IEEE Trans. Inf. Th. **14(3)** (1968), 462–467.

11. I. Csiszár, *Information-type measures of difference of probability distributions and indirect observations*, Studia Math. Hungarica **2** (1967), 299–318.

12. I. Csiszár, *On topological properties of f-divergences*, Studia Math. Hungarica **2** (1967), 329–339.

13. I. Csiszár, *A note on Jensen's inequality*, Studia Sci. Math. Hung. **1** (1966), 185–188.

14. I. Csiszár, *On topological properties of f−divergences*, Studia Math. Hungarica **2** (1967), 329–339.

15. I. Csiszár, *A note on Jensen's inequality*, Studia Sci. Math. Hung. **1** (1966), 185-188..

16. I. Csiszár and J. Körner, *Information Theory: Coding Theorem for Discrete Memoryless Systems*, Academic Press, New York, 1981.

17. D. Dacunha-Castelle, *Ecole d' ete de Probability de Saint-Flour, III-1977*, Springer Verlag, Berlin, Heidelberg, 1978.

18. S. S. Dragomir, V. Gluščević and C. E. M. Pearce, *Csiszar f-divergence, Ostrowski inequality and mutual information*, submitted.

19. S. S. Dragomir, V. Gluščević and C. E. M. Pearce, *The approximation of Csiszár f-Divergence for Absolutely Continuous Mappings*, submitted.

20. S. S. Dragomir and S. Mabizela, *Some error estimates in the trapezoidal quadrature rule*, RGMIA Research Report Collect. **2(5)** (1999), Article 6 (http://rgmia.vu.edu.au/v2n5.html).

21. S. S. Dragomir and S. Wang, *A new inequality of Ostrowski-Grüss type and its applications to the estimation of error bounds for special means and for some numerical quadrature rules*, Comp. Math. Appl. **33(11)** (1997), 15–20.

22. D. V. Gokhale and S. Kullback, *Information in Contingency Tables*, New York, Marcel Dekker, 1978.

23. J. H. Havrda and F. Charvat, *Quantification method classification process: concept of structural α-entropy*, Kybernetika **3** (1967), 30–35.

24. E. Hellinger, *Neue Bergrüirdung du Theorie quadratisher Formerus von uneudlichvieleu Veränderlicher*, J. für reine and Augeur. Math. **36** (1909), 210–271.

25. H. Jeffreys, *An invariant form for the prior probability in estimating problems*, Proc. Roy. Soc. London **186A** (1946), 453–461.

26. T. T. Kadota and L.A. Shepp, *On the best finite set of linear observables for discriminating two Gaussian signals*, IEEE Trans. Inf. Th. **13** (1967), 288–294.

27. T. Kailath, *The divergence and Bhattacharyya distance measures in signal selection*, IEEE Trans. Comm. Technology. **COM-15** (1967), 52–60.

28. J. N. Kapur, *A comparative assessment of various measures of directed divergence*, Advances in Management Studies **3** (1984), 1–16.

29. D. Kazakos and T. Cotsidas, *A decision theory approach to the approximation of discrete probability densities*, IEEE Trans. Perform. Anal. Machine Intell. **1** (1980), 61–67.

30. J. H. B. Kemperman, *On the optimum note of transmitting information*, Ann. Math. Statist. **40** (1969), 2158–2177.

31. C. Kraft, *Some conditions for consistency and uniform consistency of statistical procedures*, Univ. of California Pub. in Statistics **1** (1955), 125–142.

32. S. Kullback, *A lower bound for discrimination information in terms of variation*, IEEE Trans. Inf. Th. **13** (1967), 126–127.

33. S. Kullback, *Correction to a lower bound for discrimination information in terms of variation*, IEEE Trans. Inf. Th. **16** (1970), 771–773.

34. S. Kullback and R. A. Leibler, *On information and sufficiency*, Ann. Math. Stat. **22** (1951), 79–86.

35. S. Kullback and R. A. Leibler, *On information and sufficiency*, Annals Math. Statist. **22** (1951), 79–86.

36. L. Lecam, *Asymptotic Methods in Statistical Decision Theory*, Springer Verlag, New York, 1986.

37. J. Lin, *Divergence measures based on the Shannon entropy*, IEEE Trans. Inf. Th. **37(1)** (1991), 145–151.

38. J. Lin and S. K. M. Wong, *A new directed divergence measure and its characterization*, Int. J. General Systems **17** (1990), 73–81.

39. H. P. Mckean, Jr., *Speed of approach to equilibrium for Koc's caricature of a Maximilian gas*, Arch. Ration. Mech. Anal. **21** (1966), 343-367.

40. M. Mei, *The theory of genetic distance and evaluation of human races*, Japan J. Human Genetics **23** (1978), 341–369.

41. C. E. M. Pearce, J. Pečarić, N. Ujević and S. Varošanec, *Generaliasations of some inequalities of Ostrowski-Grüss type*, Math. Ineq. & Appl. **3(1)** (2000), 25–34.

42. E. C. Pielou, *Ecological Diversity*, Wiley, New York, 1975.

43. M. S. Pinsker, *Information and Information Stability of Random variables and processes, (in Russian)*, Moscow: Izv. Akad. Nouk, 1960.

44. C. R. Rao, *Diversity and dissimilarity coefficients: a unified approach*, Theoretic Population Biology **21** (1982), 24–43.

45. A. RÉnyi, *On measures of entropy and information*, Proc. Fourth Berkeley Symp. Math. Stat. and Prob., University of California Press **1** (1961), 547–561.

46. A. Sen, *On Economic Inequality*, Oxford University Press, London, 1973.

47. B. D. Sharma and D. P. Mittal, *New non-additive measures of relative information*, Journ. Comb. Inf. Sys. Sci. **2(4)** (1977), 122–132.

48. H. ShioyA and T. Da-te, *A generalisation of Lin divergence and the derivative of a new information divergence*, Elec. and Comm. in Japan **787** (1995), 37–40.

49. I. J. Tansja, *Generalised Information Measures and their Applications*, (http://www.mtm. ufsc.brtaneja/bhtml/bhtml.html).

50. H. Theil, *Economics and Information Theory*, North-Holland, Amsterdam, 1967.

51. H. Theil, *Statistical Decomposition Analysis*, North-Holland, Amsterdam, 1972.

52. F. Topsoe, *Some inequalities for information divergence and related measures of discrimination*, Res. Rep. Coll., RGMIA **2 (1)** (1999), 85–98.

53. G. T. Toussaint, *Sharper lower bounds for discrimination in terms of variation*, IEEE Trans. Inf. Th. **21** (1975), 99–100.

54. I. Vajda, *Note on discrimination information and variation*, IEEE Trans. Inf. Th. **16** (1970), 771–773.
55. I. Vajda, *Theory of Statistical Inference and Information*, Dordrecht-Boston, Kluwer Academic Publishers, 1989.
56. V. A. Volkonski and J. A. Rozanov, *Some limit theorems for random function-I*, (*English Trans.*), Theory Prob. Appl., (USSR) **4** (1959), 178–197.

SOME ELEMENTARY INEQUALITIES FOR THE EXPECTATION AND VARIANCE OF A RANDOM VARIABLE WHOSE PDF IS DEFINED ON A FINITE INTERVAL

N. S. BARNETT[1] AND S. S. DRAGOMIR[1]

ABSTRACT. Some elementary inequalities for the expectation and variance of a continuous random variable whose pdf is defined on a finite interval are obtained using some standard and recent results from the theory of inequalities.

1. Introduction

Let X be a continuous random variable having the probability density function f defined on a finite interval $[a, b]$.

By definition

$$E(X) := \int_a^b t f(t)\, dt$$

the *expectation* of X, and

$$\sigma^2(X) := \int_a^b (t - E(X))^2 f(t)\, dt = \int_a^b t^2 f(t)\, dt - [E(X)]^2$$

the *variance* of X.

Using some tools from the theory of inequalities, namely, Hölder's inequality, a pre-Grüss inequality, a pre-Chebychev inequality, a pre-Lupaş and Taylor's formula with integral remainder, we point out some elementary inequalities for the expectation and variance.

Received March 14, 2001.

2000 Mathematics Subject Classification. Primary 60E15; Secondary 26D15.

Key Words and Phrases. Random variable, expectation, variance.

[1] School of Communications and Informatics, Victoria University of Technology, PO Box 14428, Melbourne, MC 8001, Victoria, Australia. Email addresses: Neil.Barnett@vu.edu.au and Sever.Dragomir@vu.edu.au   http://melba.vu.edu.au/SSDragomirWeb.html

2. The Main Results

Theorem 1. *Let X be a continuous random variable defined on $[a,b]$ having p.d.f., f. Then*

(i) *we have the inequality*

$$(2.1) \qquad 0 \leq \sigma(X) \leq [b - E(X)]^{\frac{1}{2}} [E(X) - a]^{\frac{1}{2}} \leq \frac{1}{2}(b-a)$$

and

$$(2.2) \qquad
\begin{aligned}
&0 \leq [b - E(X)][E(X) - a] - \sigma^2(X) \\[2mm]
&\leq
\begin{cases}
\dfrac{(b-a)^3}{6}\|f\|_\infty, \\[4mm]
[B(q+1, q+1)]^{\frac{1}{q}}(b-a)^{2+\frac{1}{q}}\|f\|_p \\[2mm]
\left(f \in L_p[a,b]\, , \, p > 1,\ \dfrac{1}{p} + \dfrac{1}{q} = 1\right),
\end{cases}
\end{aligned}$$

where $B(\cdot, \cdot)$ is Euler's Beta function.

(ii) *If $m \leq f \leq M$ a.e. on $[a,b]$, then*

$$(2.3) \qquad \frac{m(b-a)^3}{6} \leq [b - E(X)][E(X) - a] - \sigma^2(X) \leq \frac{M(b-a)^3}{6},$$

$$(2.4) \qquad \left| [b - E(X)][E(X) - a] - \sigma^2(X) - \frac{(b-a)^2}{6} \right| \leq \frac{\sqrt{5}(b-a)^3(M-m)}{60}.$$

Proof. Note that:

$$
\begin{aligned}
\int_a^b &(b-t)(t-a) f(t)\, dt \\
&= \int_a^b [(b - E(X)) + (E(X) - t)][(E(X) - a) + (t - E(X))] f(t)\, dt \\
&= (b - E(X))(E(X) - a) \int_a^b f(t)\, dt \\
&\quad + (E(X) - a) \int_a^b (E(X) - t) f(t)\, dt \\
&\quad + (b - E(X)) \int_a^b (t - E(X)) f(t)\, dt - \int_a^b (t - E(X))^2 f(t)\, dt \\
&= [b - E(X)][E(X) - a] - \sigma^2(X)
\end{aligned}
\tag{2.5}
$$

since

$$\int_a^b f(t)\,dt = 1, \qquad \int_a^b (t - E(X))\,f(t)\,dt = 0.$$

(i) Using the fact that

$$\int_a^b (t - a)(b - t)\,f(t)\,dt \geq 0,$$

it follows that

$$\sigma^2(X) \leq [b - E(X)][E(X) - a]$$

and so the first inequality in (2.1) is established.

The second inequality in (2.1) follows from the elementary result that

$$\alpha\beta \leq \frac{1}{4}(\alpha + \beta)^2, \qquad \alpha, \beta \in \mathbb{R},$$

where $\alpha = b - E(X)$ and $\beta = E(X) - a$. The first inequality in (2.2) follows since

$$\int_a^b (t - a)(b - t)\,f(t)\,dt \leq \|f\|_\infty \int_a^b (t - a)(b - t)\,dt$$

$$= \frac{(b - a)^3}{6}\,\|f\|_\infty.$$

The second inequality is obvious by Hölder's integral inequality,

$$\int_a^b (t - a)(b - t)\,f(t)\,dt \leq \left(\int_a^b f^p(t)\,dt\right)^{\frac{1}{p}} \left(\int_a^b (t - a)^q (b - t)^q\,dt\right)^{\frac{1}{q}}$$

$$= \|f\|_p\,(b - a)^{2 + \frac{1}{q}}\,[B(q + 1, q + 1)]^{\frac{1}{q}}.$$

(ii) The inequality (2.3) is obvious, taking into account that, if $m \leq f \leq M$ a.e. on $[a, b]$, then $m(t - a)(b - t) \leq (t - a)(b - t)\,f(t) \leq M(t - a)(b - t)$ a.e. on $[a, b]$ and by integrating over $[a, b]$.

To prove (2.4), we use the following *"pre-Grüss"* inequality established in [3]:

$$(2.6) \qquad \left|\frac{1}{b - a}\int_a^b h(t)\,g(t)\,dt - \frac{1}{b - a}\int_a^b h(t)\,dt \cdot \frac{1}{b - a}\int_a^b g(t)\,dt\right|$$

$$\leq \frac{1}{2}(\phi - \gamma)\left[\frac{1}{b - a}\int_a^b g^2(t)\,dt - \left(\frac{1}{b - a}\int_a^b g(t)\,dt\right)^2\right]^{\frac{1}{2}}$$

provided that the mappings $h, g : [a, b] \to \mathbb{R}$ are measurable, all the integrals involved in (2.6) exist and are finite and $\gamma \leq h \leq \phi$ a.e. on $[a, b]$.

34   N. S. BARNETT AND S. S. DRAGOMIR

Choose, in (2.6), $h(t) = f(t)$ and $g(t) = (t-a)(b-t)$. Then

$$
\begin{aligned}
&\left| \frac{1}{b-a} \int_a^b (t-a)(b-t) f(t)\, dt \right.\\
&\qquad \left. - \frac{1}{b-a} \int_a^b (t-a)(b-t)\, dt \cdot \frac{1}{b-a} \int_a^b f(t)\, dt \right|\\
&\quad \leq \frac{1}{2}(M-m) \left[\frac{1}{b-a} \int_a^b (t-a)^2 (b-t)^2\, dt \right.\\
&\qquad \left. - \left(\frac{1}{b-a} \int_a^b (t-a)(b-t)\, dt \right)^2 \right]^{\frac{1}{2}}.
\end{aligned}
\tag{2.7}
$$

However, we have

$$
\int_a^b (t-a)(b-t)\, dt = \frac{(b-a)^3}{6}, \qquad \int_a^b f(t)\, dt = 1,
$$

$$
\int_a^b (t-a)^2 (b-t)^2\, dt
$$

$$
= (b-a)^5 \int_0^1 t^2 (1-t)^2\, dt = \frac{(b-a)^5}{30}
$$

and

$$
\frac{1}{b-a} \int_a^b (t-a)^2 (b-t)^2\, dt - \left(\frac{1}{b-a} \int_a^b (t-a)(b-t)\, dt \right)^2
$$

$$
= \frac{(b-a)^4}{30} - \frac{(b-a)^4}{36} = \frac{(b-a)^4}{180}.
$$

Consequently, by (2.7), we deduce that

$$
\left| \int_a^b (t-a)(b-t) f(t)\, dt - \frac{(b-a)^2}{6} \right| \leq \frac{1}{2}(b-a)(M-m) \left[\frac{(b-a)^4}{180} \right]^{\frac{1}{2}}
$$

$$
= \frac{(b-a)^3 (M-m)}{12\sqrt{5}}.
$$

Using (2.5), we deduce (2.4).

Remark 1. For a different proof of the inequality (2.1), see [2].

With additional information about the derivative of f, we can state the following result which complements (2.4):

Theorem 2. *Assume that the p.d.f. of X is absolutely continuous on $[a, b]$.*
   (i) *If $f' \in L_\infty [a, b]$, then we have*

$$(2.8) \quad \left| [b - E(X)][E(X) - a] - \sigma^2(X) - \frac{(b-a)^2}{6} \right| \leq \frac{\sqrt{30}}{720} \|f'\|_\infty (b-a)^3 .$$

   (ii) *If $f' \in L_2 [a, b]$, then we have*

$$(2.9) \quad \left| [b - E(X)][E(X) - a] - \sigma^2(X) - \frac{(b-a)^2}{6} \right| \leq \frac{\sqrt{5}}{60\pi} \|f'\|_2 (b-a)^3 .$$

Proof. (i) Use is made of the following "pre-Chebychev" inequality proved in [3],

$$(2.10) \quad \left| \frac{1}{b-a} \int_a^b h(t) g(t)\, dt - \frac{1}{b-a} \int_a^b h(t)\, dt \cdot \frac{1}{b-a} \int_a^b g(t)\, dt \right|$$
$$\leq \frac{1}{2\sqrt{3}} \|h'\|_\infty \left[\frac{1}{b-a} \int_a^b g^2(t)\, dt - \left(\frac{1}{b-a} \int_a^b g(t)\, dt \right)^2 \right]^{\frac{1}{2}}$$

provided that $h, g : [a, b] \to \mathbb{R}$ are measurable on $[a, b]$, the integrals involved in (2.10) exist and are finite, h is absolutely continuous and $h' \in L_\infty [a, b]$.
   Now, if we choose $h(t) = f(t)$, $g(t) = (t - a)(b - t)$ in (2.10), we get

$$\left| \int_a^b (t-a)(b-t) f(t)\, dt - \frac{(b-a)^2}{6} \right| \leq \frac{\|h'\|_\infty (b-a)}{2\sqrt{3}} \cdot \frac{(b-a)^2}{12\sqrt{5}}$$
$$= \frac{(b-a)^3 \|h'\|_\infty}{24\sqrt{30}} .$$

Using (2.5), we deduce (2.8).
   (ii) For the second part of the theorem, we use the following "pre-Lupaş" inequality as stated in [3]

$$(2.11) \quad \left| \frac{1}{b-a} \int_a^b h(t) g(t)\, dt - \frac{1}{b-a} \int_a^b h(t)\, dt \cdot \frac{1}{b-a} \int_a^b g(t)\, dt \right|$$
$$\leq \frac{b-a}{\pi} \|h'\|_2 \left[\frac{1}{b-a} \int_a^b g^2(t)\, dt - \left(\frac{1}{b-a} \int_a^b g(t)\, dt \right)^2 \right]^{\frac{1}{2}}$$

provided that g, h are as above and $h' \in L_2 [a, b]$. Now, if we choose, in (2.11), $h(t) = f(t)$, $g(t) = (t - a)(b - t)$, we obtain the desired inequality (2.9). The details are omitted.

Theorem 3. *Let X be a random variable and $f : [a,b] \to \mathbb{R}$ its p.d.f. If f is such that $f^{(n)}$ $(n \geq 0)$ is absolutely continuous on $[a,b]$, then we have the inequality*

$$\left| [E(X) - a][b - E(X)] - \sigma^2(X) - \sum_{k=0}^{n} \frac{(k+1)(b-a)^{k+3} f^{(k)}(a)}{(k+3)!} \right|$$

$$(2.12) \qquad \leq \begin{cases} \dfrac{\left\| f^{(n+1)} \right\|_\infty}{(n+1)!(n+3)(n+4)} (b-a)^{n+4} & \text{if } f^{(n+1)} \in L_\infty[a,b], \\[2ex] \dfrac{\left\| f^{(n+1)} \right\|_p (b-a)^{n+3+\frac{1}{q}}}{n!(nq+1)^{\frac{1}{q}}\left(n+2+\frac{1}{q}\right)\left(n+3+\frac{1}{q}\right)} & \text{if } f^{(n+1)} \in L_p[a,b], \, p > 1, \\[2ex] \dfrac{\left\| f^{(n+1)} \right\|_1 (b-a)^{n+3}}{n!(n+2)(n+3)} & \text{if } f^{(n+1)} \in L_1[a,b], \end{cases}$$

where $\|\cdot\|_p$ $(1 \leq p \leq \infty)$ are the usual Lebesgue norms on $[a,b]$, i.e.,

$$\|g\|_\infty := \operatorname*{ess\,sup}_{t \in [a,b]} |g(t)|, \quad \|g\|_p := \left(\int_a^b |g(t)|^p \, dt \right)^{\frac{1}{p}} \quad p \geq 1.$$

Proof. The following Taylor's formula with integral remainder is well known in the literature (see for example [1]):

$$(2.13) \qquad f(t) = \sum_{k=0}^{n} \frac{(t-a)^k}{k!} f^{(k)}(a) + \frac{1}{n!} \int_a^t (t-s)^n f^{(n+1)}(s) \, ds$$

for all $t \in [a,b]$. Since

$$(2.14) \qquad [E(X) - a][b - E(X)] - \sigma^2(X) = \int_a^b (t-a)(b-t) f(t) \, dt,$$

then we have

$$[E(X) - a][b - E(X)] - \sigma^2(X)$$

$$= \int_a^b (t-a)(b-t) \left[\sum_{k=0}^{n} \frac{(t-a)^k}{k!} f^{(k)}(a) \right.$$

$$(2.15) \qquad \left. + \frac{1}{n!} \int_a^t (t-s)^n f^{(n+1)}(s) \, ds \right] dt$$

$$= \sum_{k=0}^{n} \frac{f^{(k)}(a)}{k!} \int_a^b (t-a)^{k+1} (b-t) \, dt$$

$$+ \frac{1}{n!} \int_a^b \left[(t-a)(b-t) \int_a^t (t-s)^n f^{(n+1)}(s) \, ds \right] dt.$$

Using the transform, $t = (1 - u)\, a + ub$, we have

$$\int_a^b (t - a)^{k+1} (b - t)\, dt = (b - a)^{k+3} \int_0^1 u^{k+1} (1 - u)\, du$$

$$= \frac{1}{(k + 2)(k + 3)}$$

and, by (2.15), we deduce that

$$\left| [E(X) - a]\, [b - E(X)] - \sigma^2(X) - \sum_{k=0}^n \frac{(k + 1)(b - a)^{k+3} f^{(k)}(a)}{(k + 3)!} \right|$$

$$\leq \frac{1}{n!} \int_a^b (t - a)(b - t) \left| \int_a^t (t - s)^n f^{(n+1)}(s)\, ds \right| dt =: M(a, b).$$

However, for all $t \in [a, b]$, we have

$$\left| \int_a^t (t - s)^n f^{(n+1)}(s)\, ds \right| \leq \int_a^t |t - s|^n \left| f^{(n+1)}(s) \right| ds$$

$$\leq \sup_{s \in [a,b]} \left| f^{(n+1)}(s) \right| \int_a^t (t - s)^n\, ds$$

$$\leq \left\| f^{(n+1)} \right\|_\infty \frac{(t - a)^{n+1}}{n + 1}.$$

By Hölder's integral inequality, we have,

$$\left| \int_a^t (t - s)^n f^{(n+1)}(s)\, ds \right| \leq \left(\int_a^t \left| f^{(n+1)}(s) \right|^p ds \right)^{\frac{1}{p}} \left(\int_a^t (t - s)^{nq}\, ds \right)^{\frac{1}{q}}$$

$$\leq \left\| f^{(n+1)} \right\|_p \left[\frac{(t - a)^{nq+1}}{nq + 1} \right]^{\frac{1}{q}}, \quad \frac{1}{p} + \frac{1}{q} = 1,\ p > 1,$$

for all $t \in [a, b]$. Finally, we observe that

$$\left| \int_a^t (t - s)^n f^{(n+1)}(s)\, ds \right| \leq \int_a^t (t - s)^n \left| f^{(n+1)}(s) \right| ds$$

$$\leq (t - a)^n \int_a^t \left| f^{(n+1)}(s) \right| ds$$

$$\leq (t - a)^n \left\| f^{(n+1)} \right\|_1$$

for all $t \in [a, b]$. Consequently, we have

$$
M(a, b) \leq \frac{1}{n!} \times
\begin{cases}
\dfrac{\left\| f^{(n+1)} \right\|_\infty}{n+1} \displaystyle\int_a^b (t-a)^{n+2} (b-t)\, dt, \\[2ex]
\dfrac{\left\| f^{(n+1)} \right\|_p}{(nq+1)^{\frac{1}{q}}} \displaystyle\int_a^b (t-a)^{n+1+\frac{1}{q}} (b-t)\, dt, \\[2ex]
\left\| f^{(n+1)} \right\|_1 \displaystyle\int_a^b (t-a)^{n+1} (b-t)\, dt
\end{cases}
$$

$$
=
\begin{cases}
\dfrac{\left\| f^{(n+1)} \right\|_\infty}{n+1} (b-a)^{n+4} \displaystyle\int_0^1 u^{n+2} (1-u)\, du, \\[2ex]
\dfrac{\left\| f^{(n+1)} \right\|_p}{(nq+1)^{\frac{1}{q}}} (b-a)^{n+3+\frac{1}{q}} \displaystyle\int_0^1 u^{n+1+\frac{1}{q}} (1-u)\, du, \\[2ex]
\left\| f^{(n+1)} \right\|_1 (b-a)^{n+3} \displaystyle\int_0^1 u^{n+1} (1-u)\, du
\end{cases}
$$

and, as

$$
\int_0^1 u^{n+2} (1-u)\, du = \frac{1}{(n+3)(n+4)},
$$

$$
\int_0^1 u^{n+1+\frac{1}{q}} (1-u)\, du = \frac{1}{\left(n+2+\frac{1}{q}\right)\left(n+3+\frac{1}{q}\right)},
$$

$$
\int_0^1 u^{n+1} (1-u)\, du = \frac{1}{(n+2)(n+3)},
$$

the inequality (2.12) is proved.

Remark 2. A similar result can be obtained if use is made of a Taylor expansion around the point b.

REFERENCES

1. T. M. Apostol, *Mathematical Analysis*, 2nd Edition, Addison Wesley Publishing Company, 1975.
2. N. S. Barnett, P. Cerone, S. S. Dragomir and J. Roumeliotis, *Some inequalities for the dispertion of a random variable whose pdf is defined on a finite interval*, J. Inequal. Pure and Appl. math. **2(1)** (2000), Article 1, http://jipam.vu.edu.au.
3. M. Matić, J. E. Pečarić and N. Uhević, *On new estimation of the remainder in generalized Taylor's formula*, Mathematical Inequalities and Applications **2(3)** (1999), 343–361.

SOME LIMIT THEOREMS ON THE INCREMENTS OF l^p-VALUED GAUSSIAN PROCESSES

YONG KAB CHOI[1] AND SEAUNG HYUNE LEE[1]

Dedicated to Professor Norio Kôno on the occasion of his 60th birthday

ABSTRACT. In this paper, we establish some limit theorems on the increments of an l^p-valued Gaussian process under weaker conditions than those of some theorems in Csörgő and Shao [6].

1. Introduction and Results

Limit theorems on the increments of Wiener processes and Gaussian processes are deeply related to the properties of their sample paths. So the convergence properties of increments of Wiener processes and Gaussian processes have attracted the attention of many probabilists in a last few decades. Recently, the limit theorems on the increments of several l^p-valued Gaussian processes have been studied in various directions by many authors: for instance, Csörgő and Shao [6], Csáki, Csörgő and Shao ([4], [5]), Csáki and Csörgő [3], Lin [9], Shao [10] and etc. For our purpose we shall quote here some results in [6]. Let $\{Y(t), t \geq 0\} = \{X_k(t), t \geq 0\}_{k=1}^{\infty}$ be a sequence of independent continuous Gaussian processes with $EX_k(t) = 0$ and stationary increments

$$\sigma_k^2(h) = E\{X_k(t+h) - X_k(t)\}^2,$$

Received November 9, 2000.

1991 *Mathematics Subject Classification.* 60F15, 60G15.

Key Words and Phrases. L^p-norm, Wiener process, Gaussian process, Borel-Cantelli lemma.

This work was supported by grant No. 2001-1-10400-010-2 from the Basic Research Program of the Korea Science & Engineering Foundation.

[1] Department of Mathematics, Gyeongsang National University, Chinju 660-701, Korea. Email address: mathykc@nongae.gsnu.ac.kr and matht8@hanmail.net

where $\sigma_k(h)$ are nondecreasing continuous functions of $h > 0$. Put

$$
(1.1) \qquad \sigma(p,h) = \left(\sum_{k=1}^{\infty} \sigma_k^p(h) \right)^{1/p}, \quad 1 \leq p < \infty,
$$

$$
(1.2) \qquad \sigma^*(h) = \max_{k \geq 1} \sigma_k(h),
$$

$$
(1.3) \qquad \tilde{\sigma}(p,h) = \begin{cases} \sigma(\frac{2p}{2-p}, h), & \text{if } 1 \leq p < 2, \\ \sigma^*(h), & \text{if } 2 \leq p < \infty, \end{cases}
$$

$$
\delta_p^p := E|N(0,1)|^p = \frac{2^{p/2}}{\sqrt{\pi}} \int_0^{\infty} x^{(p-1)/2} e^{-x} \, dx, \quad 1 \leq p < \infty.
$$

Csörgő and Shao proved the following results (cf. Theorem 3.2 and Theorem 3.3 in [6]):

Theorem A. *Let a_T $(0 < T < \infty)$ be a positive continuous function of T and b_T be a nonnegative continuous function of T. Put $a^* = \sup_{T>0} a_T$. Assume that $\sigma(p,h)/h^\alpha$ and $\tilde{\sigma}(p,h)/h^\alpha$ are quasi-increasing on $(0, a^*)$ for some $\alpha > 0$ and that*

$$
(i) \qquad \frac{b_T}{a_T} + a_T + \frac{1}{a_T} \to \infty \quad \text{as } T \to \infty.
$$

Then we have

$$
\limsup_{T \to \infty} \sup_{0 \leq t \leq b_T} \sup_{0 \leq s \leq a_T} \beta(p,T) \| Y(t+s) - Y(t) \|_{l^p} \leq 1 \quad \text{a.s.,}
$$

where

$$
\beta(p,T)^{-1} = \delta_p \sigma(p, a_T) + \tilde{\sigma}(p, a_T)(2(\log(b_T/a_T) + \log\log(a_T + 1/a_T)))^{1/2}.
$$

Theorem B. *Let a_T and b_T be positive continuous functions of T. Assume that $\sigma(p,h) < \infty$, as well as*

$$
(ii) \qquad \frac{\log(b_T/a_T)}{\log\log(a_T + \frac{1}{a_T})} \to \infty \quad \text{as } T \to \infty
$$

and

$$
(iii) \qquad \limsup_{T \to \infty} \max_{(b_T/a_T)^\epsilon \leq j \leq b_T/a_T} \max_{k \geq 1} \{ \sigma_k^{-2}(a_T)
$$
$$
\times E[(X_k(a_T) - X_k(0))(X_k(ja_T) - X_k((j-1)a_T))] \} \leq 0
$$

for each $\varepsilon > 0$ are satisfied. Then we have

$$\liminf_{T\to\infty} \sup_{0\le t\le b_T} \sup_{0\le s\le a_T} \frac{\|Y(t+s) - Y(t)\|_{l^p}}{\tilde{\sigma}(p, a_T)\sqrt{2\log(b_T/a_T)}} \ge 1 \quad a.s.$$

The main aim of this paper is to generalize Theorems A and B under mild conditions. Throughout the paper, we shall always assume that $\sigma_k(h)$ are non-decreasing continuous and regularly varying functions of $h > 0$ with exponents α_k $(0 < \alpha_k < \alpha_{k+1} < 1, \ k = 1, 2, \cdots)$ at 0 and ∞, and we shall use the same notations as in (1.1)-(1.3). Assume that $Y(t) = (X_1(t), X_2(t), \cdots)$, $0 \le t < \infty$, is an infinite dimensional Gaussian process with l^p-norm $\|\cdot\|_{l^p}$ and independent components $X_i(t)$, $i = 1, 2, \cdots$. Let a_T and b_T be positive functions of $T > 0$ and let $\liminf_{T\to\infty} a_T > 0$. Denote

$$\beta_1(T) = \left\{2\left(\log\frac{b_T}{a_T} + \log\log\left(\tilde{\sigma}(p, a_T) + \frac{1}{\tilde{\sigma}(p, a_T)}\right)\right)\right\}^{1/2},$$

$$\beta_2(T) = \left\{2\left(\log\frac{b_T}{a_T} + \log\log\left(a_T + \frac{1}{a_T}\right)\right)\right\}^{1/2},$$

where $\log x = \ln(\max\{x, 1\})$. From (3.10) of [6], we know that

$$(1.4) \qquad \tilde{\sigma}(p, h) + \frac{1}{\tilde{\sigma}(p, h)} \le 4\left(h + \frac{1}{h}\right)\left(\tilde{\sigma}(p, 1) + \frac{1}{\tilde{\sigma}(p, 1)}\right)$$

for each $h > 0$. Thus we have $\beta_1(T) \le \beta_2(T)$ for all large T.

The main results are as follows:

Theorem 1.1. *Assume that*

$$(i) \qquad \frac{b_T}{a_T} + a_T + \frac{1}{a_T} \to \infty \qquad as \quad T \to \infty.$$

Then we have

$$(1.5) \qquad \limsup_{T\to\infty} \sup_{0\le t\le b_T} \sup_{0\le s\le a_T} \frac{\|Y(t+s) - Y(t)\|_{l^p}}{\tilde{\sigma}(p, a_T)\beta_1(T)} \le 1 \quad a.s.$$

Theorem 1.2. *Assume that*

$$(ii) \qquad \frac{\log(b_T/a_T)}{\log\log(a_T + \frac{1}{a_T})} \to \infty \quad as \ T \to \infty$$

and

(iii)′   for $h > 0$, $\sigma_k(h)$ are twice differentiable which satisfy

$$\left| \frac{d^2\sigma_k^2(h)}{dh^2} \right| \le c \frac{\sigma_k^2(h)}{h^2}$$

for some $c > 0$. Then we have

$$(1.6) \qquad \liminf_{T\to\infty} \sup_{0\le t\le b_T} \sup_{0\le s\le a_T} \frac{\|Y(t+s) - Y(t)\|_{l^p}}{\tilde{\sigma}(p, a_T)\beta_2(T)} \ge 1 \quad a.s.$$

From Theorems 1.1 and 1.2, we have the following limit theorem:

Corollary 1.1. *Under the assumptions of Theorem 1.2, we have*

$$\lim_{T\to\infty} \sup_{0\le t\le b_T} \sup_{0\le s\le a_T} \frac{\|Y(t+s) - Y(t)\|_{l^p}}{\tilde{\sigma}(p, a_T)\beta_i(T)} = 1, \quad i = 1, 2, \quad a.s.$$

2. The Proofs

Let $\mathbb{T}$ be a compact subset of $[0, \infty)^N$ with the Euclidean norm $\|\cdot\|$. Let $\{X_i(t), t \in \mathbb{T}\}$, $i = 1, 2, \cdots$, be independent separable Gaussian processes with $X_i(0) = 0$, $EX_i(t) = 0$ and $\sqrt{E\{X_i(t) - X_i(s)\}^2} = \sigma_i(\|t - s\|)$, where $\sigma_i(h)$ are nondecreasing continuous and regularly varying functions of $h > 0$ with exponents α_i ($0 < \alpha_i < \alpha_{i+1} < 1$, $i = 1, 2, \cdots$) at 0 and ∞. For $t \in \mathbb{T}$, let $Y(t) = (X_1(t), X_2(t), \cdots)$ be an infinite dimensional N-parameter Gaussian process with $\|\cdot\|_{l^p}$. Suppose that

$$(2.1) \qquad 0 < \sup_{t\in\mathbb{T}} E\{X_i(t)\}^2 =: \Gamma_i^2 < \infty, \quad i = 1, 2, \cdots,$$

and

$$(2.2) \qquad \sigma_i(h) \le \varphi_i(h),$$

where $\varphi_i(\cdot)$ are nondecreasing continuous functions. Put

$$(2.3) \qquad \Gamma = \begin{cases} \left(\sum_{i=1}^{\infty} \Gamma_i^{\frac{2p}{2-p}}\right)^{\frac{2-p}{2p}}, & \text{if } 1 \le p < 2, \\ \max_{i\ge 1} \Gamma_i, & \text{if } p \ge 2, \end{cases}$$

$$(2.4) \qquad \varphi(p, h) := \left(\sum_{i=1}^{\infty} \varphi_i^p(h)\right)^{1/p}, \quad p \ge 1,$$

$$(2.5) \qquad \varphi^*(h) = \max_{i\ge 1} \varphi_i(h),$$

$$(2.6) \qquad \tilde{\varphi}(p, h) = \begin{cases} \varphi\left(\frac{2p}{2-p}, h\right), & \text{if } 1 \le p < 2, \\ \varphi^*(h), & \text{if } p \ge 2. \end{cases}$$

Using the Borell inequality (2.6) in [1], we can obtain the following Lemma 2.1 by the same method as the proof of Lemma 3.1 in [6]. Here we omit the details.

Lemma 2.1. *Let $p \geq 1, \{\xi_n, n \geq 1\}$ be independent normal random variables with $E\xi_n = 0$ and $\sum_{i=1}^{\infty}(E\xi_i^2)^{p/(2-p)} < \infty$. Then there exists a constant $C > 0$ such that for all large x*

$$P\left\{\left(\sum_{i=1}^{\infty}|\xi_i|^p\right)^{1/p} \geq x\right\}$$
$$= \begin{cases} \exp\left(Cx - \left(x^2/2\left(\sum_{i=1}^{\infty}(E\xi_i^2)^{\frac{p}{2-p}}\right)^{\frac{2-p}{p}}\right)\right), & 1 \leq p < 2, \\ \exp\left(Cx - \left(x^2/2\max_{i\geq 1}E\xi_i^2\right)\right), & p \geq 2. \end{cases}$$

Lemma 2.2. *Let $p \geq 1$. Then there exists a constant $C > 0$ such that for all large x*

$$P\left\{\|Y(\mathbf{t}) - Y(\mathbf{s})\|_{l^p} \geq x\tilde{\sigma}(p, \|\mathbf{t} - \mathbf{s}\|)\right\} \leq \exp\left(Cx - \frac{x^2}{2}\right).$$

Proof. For $i = 1, 2, \cdots$, let $\xi_i = X_i(\mathbf{t}) - X_i(\mathbf{s})$ in Lemma 2.1. Then

$$\left(\sum_{i=1}^{\infty}|\xi_i|^p\right)^{\frac{1}{p}} = \|Y(\mathbf{t}) - Y(\mathbf{s})\|_{l^p},$$
$$E\xi_i^2 = \sigma_i^2(\|\mathbf{t} - \mathbf{s}\|), \quad \max_{i\geq 1}E\xi_i^2 = \sigma^{*2}(\|\mathbf{t} - \mathbf{s}\|),$$
$$\left(\sum_{i=1}^{\infty}(E\xi_i^2)^{\frac{p}{2-p}}\right)^{\frac{2-p}{p}} = \tilde{\sigma}^2(p, \|\mathbf{t} - \mathbf{s}\|), \quad 1 \leq p < 2.$$

Replacing x in Lemma 2.1 by $x\tilde{\sigma}(p, \|\mathbf{t} - \mathbf{s}\|)$, $p \geq 1$, it follows from Lemma 2.1 that

$$P\left\{\|Y(\mathbf{t}) - Y(\mathbf{s})\|_{l^p} \geq x\tilde{\sigma}(p, \|\mathbf{t} - \mathbf{s}\|)\right\}$$
$$= P\left\{\left(\sum_{i=1}^{\infty}|\xi_i|^p\right)^{\frac{1}{p}} \geq x\tilde{\sigma}(p, \|\mathbf{t} - \mathbf{s}\|)\right\}$$
$$\leq \exp\left(Cx - \frac{x^2}{2}\right), \quad p \geq 1.$$

The next Lemma 2.3 is another version of Fernique lemma [7] on an l^p-valued Gaussian process:

Lemma 2.3. *Let $\lambda > 0$ and $\mathcal{A} > (C + \sqrt{C^2 + N\ln 4})(2\sqrt{2} + 2)$ for some $C > 0$. Then there exists a positive constant $C_{\mathcal{A}}$ depending only on $\mathcal{A}$ such that for all large x*

$$P\left\{\sup_{t\in\mathbb{T}} \|Y(t)\|_{l^p} \geq x\left(\Gamma + \mathcal{A}\int_0^\infty \tilde{\varphi}(p, \sqrt{N}\lambda 2^{-y^2})\,dy\right)\right\}$$
$$\leq C_{\mathcal{A}}\frac{m(\mathbb{T})}{\lambda^N}\exp\left(Cx - \frac{x^2}{2}\right), \quad p \geq 1,$$

where $m(\mathbb{T})$ is the Lebesgue measure of $\mathbb{T}$.

Proof. For each $n = 0, 1, 2, \cdots$, set $\epsilon_n = \lambda 2^{-2^n}$, $\lambda > 0$. For any compact subset $\mathbb{T}$ of $[0, \infty)^N$, let $d(\mathbb{T})$ be a diameter of $\mathbb{T}$. Let $\left\{S_i^{(n)} : i = 1, 2, \cdots, N_{\epsilon_n}(\mathbb{T})\right\}$ be a minimal ε_n-net of $\mathbb{T}$, where

$$N_{\epsilon_n}(\mathbb{T}) = \min\left\{k : \mathbb{T} \subset \cup_{i=1}^k S_i^{(n)}, d(S_i^{(n)}) \leq \epsilon_n\right\}.$$

Then there is a positive constant c such that $N_{\epsilon_n}(\mathbb{T}) \leq c\, m(\mathbb{T})/(\epsilon_n)^N$. Let $t_i^{(n)}$ be a point in $S_i^{(n)}$ and put $\Delta_n = \bigcup_{i=1}^{N_{\epsilon_n}(\mathbb{T})}\{t_i^{(n)}\}$. Let $\mathcal{B} > C + \sqrt{C^2 + N\ln 4}$ and $\mathcal{A} = (2\sqrt{2} + 2)\mathcal{B}$. For a large number x, set

$$x_k = x\mathcal{B}\tilde{\varphi}(p, \sqrt{N}\epsilon_{k-1})2^{k/2}, \quad k \geq 1.$$

Let $\delta_k = 2^{(k-1)/2}$ for $k \geq 0$. Then $2^{k/2} = (2\sqrt{2} + 2)(\delta_k - \delta_{k-1})$. Thus we have

$$\sum_{k=1}^\infty x_k = x\mathcal{A}\sum_{k=1}^\infty \tilde{\varphi}(p, \sqrt{N}\lambda 2^{-\delta_k^2})(\delta_k - \delta_{k-1})$$
$$\leq x\mathcal{A}\sum_{k=1}^\infty \int_{\delta_{k-1}}^{\delta_k} \tilde{\varphi}(p, \sqrt{N}\lambda 2^{-y^2})dy$$
$$\leq x\mathcal{A}\int_0^\infty \tilde{\varphi}(p, \sqrt{N}\lambda 2^{-y^2})\,dy.$$

Therefore, it follows that

$$P\left\{\sup_{t\in\mathbb{T}} \|Y(t)\|_{l^p} \geq x\left(\Gamma + \mathcal{A}\int_0^\infty \tilde{\varphi}(p, \sqrt{N}\lambda 2^{-y^2})dy\right)\right\}$$
$$\leq P\left\{\sup_{t\in\mathbb{T}} \|Y(t)\|_{l^p} \geq x\Gamma + \sum_{k=1}^\infty x_k\right\}$$
$$\text{(2.7)} \qquad = P\left\{\lim_{n\to\infty}\sup_{t\in\Delta_n} \|Y(t)\|_{l^p} \geq x\Gamma + \sum_{k=1}^\infty x_k\right\}$$
$$\leq \lim_{n\to\infty} P\left\{\sup_{t\in\Delta_n} \|Y(t)\|_{l^p} \geq x\Gamma + \sum_{k=1}^n x_k\right\}.$$

Let

$$B_0 = \left\{ \sup_{t \in \Delta_0} \|Y(\mathbf{t})\|_{l^p} \geq x\Gamma \right\}$$

and

$$B_n = \left\{ \sup_{t \in \Delta_n} \|Y(\mathbf{t})\|_{l^p} \geq x\Gamma + \sum_{k=1}^{n} x_k \right\}, \quad n \geq 1.$$

By induction, we have

$$
\begin{aligned}
P(B_n) &= P(B_n \cap B_{n-1}) + P(B_n \cap B_{n-1}^c) \\
&\leq P(B_{n-1}) + P(B_n \cap B_{n-1}^c) \\
&\leq P(B_{n-2}) + P(B_{n-1} \cap B_{n-2}^c) + P(B_n \cap B_{n-1}^c) \\
&\leq P(B_0) + \sum_{k=1}^{\infty} P(B_k \cap B_{k-1}^c).
\end{aligned}
$$

(2.8)

For $n \geq 1$, we have

$$
\begin{aligned}
&P(B_n \cap B_{n-1}^c) \\
&= P\left\{ \left(\sup_{t \in \Delta_n} \|Y(\mathbf{t})\|_{l^p} \geq x\Gamma + \sum_{k=1}^{n} x_k \right) \right. \\
&\qquad \left. \bigcap \left(\sup_{s \in \Delta_{n-1}} \|Y(\mathbf{s})\|_{l^p} < x\Gamma + \sum_{k=1}^{n-1} x_k \right) \right\} \\
&\leq P\left\{ \bigcup_{t \in \Delta_n} \left\{ \|Y(\mathbf{t})\|_{l^p} \geq x\Gamma + \sum_{k=1}^{n} x_k \right\} \bigcap \bigcap_{s \in \Delta_{n-1}} \left\{ \|Y(\mathbf{s})\|_{l^p} < x\Gamma + \sum_{k=1}^{n-1} x_k \right\} \right\} \\
&\leq P\left\{ \bigcup_{t \in \Delta_n} \bigcup_{\substack{s \in \Delta_{n-1} \\ \|\mathbf{t}-\mathbf{s}\| \leq \sqrt{N}\epsilon_{n-1}}} \left\{ \|Y(\mathbf{t})\|_{l^p} - \|Y(\mathbf{s})\|_{l^p} \geq x_n \right\} \right\} \\
&\leq \sum_{t \in \Delta_n} \sum_{\substack{s \in \Delta_{n-1} \\ \|\mathbf{t}-\mathbf{s}\| \leq \sqrt{N}\epsilon_{n-1}}} P\left\{ \|Y(\mathbf{t}) - Y(\mathbf{s})\|_{l^p} \geq x_n \right\}.
\end{aligned}
$$

It follows from (2.2), (2.4) and Lemma 2.2 that

$$
\begin{aligned}
&P(B_n \cap B_{n-1}^c) \\
&\leq \sum_{t \in \Delta_n} \sum_{\substack{s \in \Delta_{n-1} \\ \|\mathbf{t}-\mathbf{s}\| \leq \sqrt{N}\epsilon_{n-1}}} P\left\{ \|Y(\mathbf{t}) - Y(\mathbf{s})\|_{l^p} \geq x_n \right\}
\end{aligned}
$$

$$\leq \sum_{\mathbf{t}\in\Delta_n} \sum_{\substack{\mathbf{s}\in\Delta_{n-1}\\ \|\mathbf{t}-\mathbf{s}\|\leq\sqrt{N}\epsilon_{n-1}}} P\{\|Y(\mathbf{t})-Y(\mathbf{s})\|_{l^p}\geq x\mathcal{B}\tilde{\sigma}(p,\|\mathbf{t}-\mathbf{s}\|)2^{n/2}\}$$

$$\leq c\frac{2^{N2^n}m(\mathbb{T})}{\lambda^N}\exp\left(Cx\mathcal{B}2^{n/2}-\frac{1}{2}x^2\mathcal{B}^2 2^n\right)$$

and hence

$$\sum_{n=1}^{\infty} P(B_n\cap B_{n-1}^c)$$

$$(2.9) \qquad \leq \sum_{n=1}^{\infty} c\frac{m(\mathbb{T})}{\lambda^N}\exp\left(-2^n\left(\frac{\mathcal{B}^2}{2}-C\mathcal{B}2^{-n/2}-N\ln 2\right)\right)\exp\left(-\frac{x^2}{2}\right)$$

$$\leq C_{\mathcal{A}}\frac{m(\mathbb{T})}{\lambda^N}\exp\left(-\frac{x^2}{2}\right).$$

On the other hand, we have

$$P(B_0)=P\left\{\sup_{\mathbf{t}\in\Delta_0}\|Y(\mathbf{t})\|_{l^p}\geq x\Gamma\right\}$$

$$(2.10) \qquad\qquad \leq c\frac{m(\mathbb{T})}{\epsilon_0^N}\exp\left(Cx-\frac{x^2}{2}\right)$$

$$= c2^N\frac{m(\mathbb{T})}{\lambda^N}\exp\left(Cx-\frac{x^2}{2}\right).$$

From $(2.7)\sim(2.10)$, the proof is completed.

Using Lemma 2.3, we can estimate an upper bound of the following large deviation probabilities:

Lemma 2.4. *Let a_T and b_T be positive functions of T. Then, for any $\varepsilon > 0$ there exists a positive constant C_ε depending only on ε such that*

$$P\left\{\sup_{0\leq t\leq b_T}\sup_{0\leq s\leq a_T}\frac{\|Y(t+s)-Y(t)\|_{l^p}}{\tilde{\sigma}(p,a_T)}\geq x\right\}$$

$$\leq C_\varepsilon\left(\frac{b_T}{a_T}\right)\exp\left(Cx-\frac{x^2}{2+\varepsilon}\right),$$

for all large $x > 0$, where $C > 0$ is a constant and $p \geq 1$.

Proof. Let $\mathbb{D}=\{(t,s):0\leq t\leq b_T,\ 0\leq s\leq a_T\}$ be a two-parameter set. In order to apply Lemma 2.3, we set, for $i\geq 1$,

$$Z_i(t,s)=\frac{X_i(t+s)-X_i(t)}{\tilde{\sigma}(p,a_T)}\ ,\quad (t,s)\in\mathbb{D},$$

and

$$\varphi_i(z) = \frac{2\sigma_i(\sqrt{2}z)}{\tilde{\sigma}(p, a_T)} , \quad z > 0.$$

Clearly, for $i \geq 1$, we have $E\{Z_i(t, s)\} = 0$, $\Gamma_i = \sigma_i(a_T)/\tilde{\sigma}(p, a_T)$ and $\Gamma = 1$ from (2.3). Letting $\mathbf{u} = (t', s')$ and $\mathbf{v} = (t'', s'')$, it follows that

$$\begin{aligned}
E\{Y_i(\mathbf{u}) &- Y_i(\mathbf{v})\}^2 \\
&\leq \frac{2}{\tilde{\sigma}(p, a_T)} \{\sigma_i^2(|(t' + s') - (t'' + s'')|) + \sigma_i^2(|t' - t''|)\} \\
&\leq \frac{4}{\tilde{\sigma}(p, a_T)} \sigma_i^2(\sqrt{2}\sqrt{|t' - t''|^2 + |s' - s''|^2}) \\
&= \varphi_i^2(\|\mathbf{u} - \mathbf{v}\|).
\end{aligned}$$

On the other hand, let us prove that, for any $\varepsilon > 0$, there exists $c_\varepsilon > 0$ such that

$$(2.11) \qquad A \int_0^\infty \tilde{\varphi}(p, \sqrt{2}\, c_\varepsilon\, a_T\, 2^{-y^2})\, dy < \varepsilon/8$$

for all $T > 0$, where $A > (C + \sqrt{C^2 + N \ln 4})(2\sqrt{2} + 2)$ for some $C > 0$. If $1 \leq p < 2$, then there exists $c_\varepsilon > 0$ such that

$$\begin{aligned}
A \int_0^\infty \tilde{\varphi}(p, \sqrt{2}\, c_\varepsilon\, a_T\, 2^{-y^2})\, dy &= A \int_0^\infty \varphi\left(\frac{2p}{2-p}, \sqrt{2}\, c_\varepsilon\, a_T\, 2^{-y^2}\right) dy \\
&= 2A \int_0^\infty \frac{\sigma\left(\frac{2p}{2-p}, 2\, c_\varepsilon\, a_T\, 2^{-y^2}\right)}{\sigma\left(\frac{2p}{2-p}, a_T\right)}\, dy \\
&\leq 2A \int_0^\infty \frac{\sigma_1\left(2\, c_\varepsilon\, a_T\, 2^{-y^2}\right)}{\sigma_1(a_T)}\, dy \\
&\leq K_1\, c_\varepsilon^{\alpha_1} \int_0^\infty 2^{-\alpha_1 y^2}\, dy < \varepsilon/8
\end{aligned}$$

for all $T > 0$, where K_1 is a constant. As for $p \geq 2$, we have

$$\begin{aligned}
A \int_0^\infty \tilde{\varphi}(p, \sqrt{2}\, c_\varepsilon\, a_T\, 2^{-y^2})\, dy &= A \int_0^\infty \varphi^*(\sqrt{2}\, c_\varepsilon\, a_T\, 2^{-y^2})\, dy \\
&= 2A \int_0^\infty \frac{\sigma^*\left(2\, c_\varepsilon\, a_T\, 2^{-y^2}\right)}{\sigma^*(a_T)}\, dy \\
&\leq K_2\, c_\varepsilon^{\alpha_1} \int_0^\infty 2^{-\alpha_1 y^2}\, dy < \varepsilon/8
\end{aligned}$$

for all $T > 0$, where K_2 is a constant. Let $\{Z(t,s) = (Z_1(t,s), Z_2(t,s), \cdots),$ $(t,s) \in \mathbb{D}\}$ be an infinite dimensional two-parameter Gaussian process with $\|\cdot\|_{l^p}$. For a large number x, let $u = x/(1 + (\varepsilon/8))$. Then it follows from Lemma 2.3 that

$$P\left\{ \sup_{0 \leq t \leq \, b_T} \sup_{0 \leq s \leq a_T} \frac{\|Y(t+s) - Y(t)\|_{l^p}}{\tilde{\sigma}(p, a_T)} \geq x \right\}$$
$$\leq P\left\{ \sup_{(t,s) \in \mathbb{D}} \|Z(t,s)\|_{l^p} \geq u\left(\Gamma + \mathcal{A} \int_0^\infty \tilde{\varphi}\left(p, \sqrt{2}\, c_\varepsilon\, a_T\, 2^{-y^2}\right) dy\right)\right\}$$
$$\leq C_{\varepsilon,\mathcal{A}} \left(\frac{b_T}{a_T}\right) \exp\left(\frac{8C}{8+\varepsilon} x - \frac{x^2}{2+\varepsilon}\right),$$

where $C_{\varepsilon,\mathcal{A}}$ is a positive constant depending only on ε and $\mathcal{A}$.

Proof of Theorem 1.1. Let $\theta = \sqrt{1+\epsilon}$ for any given $\epsilon > 0$. Define

$$A_k = \{T : \theta^k \leq \tilde{\sigma}(p, a_T) \leq \theta^{k+1}\}, \quad -\infty < k < \infty,$$
$$A_{k,j} = \left\{T : 2^j \leq \frac{b_T}{a_T} \leq 2^{j+1}, \quad T \in A_k\right\}, \quad -\infty < j < \infty,$$
$$a_{T_{k,j}} = \sup\{a_T : T \in A_{k,j}\}, \quad b_{T_{k,j}} = \sup\{b_T : T \in A_{k,j}\}.$$

By (i), we have

$$\limsup_{T \to \infty} \sup_{0 \leq t \leq b_T} \sup_{0 \leq s \leq a_T} \frac{\|Y(t+s) - Y(t)\|_{l^p}}{\tilde{\sigma}(p, a_T)\beta_1(T)}$$

(2.12)
$$\leq \limsup_{\substack{|k|+|l| \to \infty \\ j \leq -\bar{l} < 0}} \sup_{\substack{j \geq l \geq 0}} \sup_{T \in A_{k,j}} \sup_{0 \leq t \leq b_T} \sup_{0 \leq s \leq a_T} \frac{\|Y(t+s) - Y(t)\|_{l^p}}{\tilde{\sigma}(p, a_T)\beta_1(T)}.$$

We proceed with the proof to divide into two cases: $j \geq l \geq 0$, $j \leq -l < 0$.

CASE 1: The case in which (2.12) holds when $j \geq l \geq 0$.
We will prove that

$$\limsup_{|k|+l \to \infty} \sup_{j \geq l} \sup_{0 \leq t \leq b_{T_{k,j}}} \sup_{0 \leq s \leq a_{T_{k,j}}} \frac{\|Y(t+s) - Y(t)\|_{l^p}}{\theta^k \left\{2\left(\log 2^j + \log\log \theta^{|k|}\right)\right\}^{1/2}}$$

(2.13)
$$\leq \theta \cdot \limsup_{|k|+l \to \infty} \sup_{j \geq l} \sup_{0 \leq t \leq b_{T_{k,j}}} \sup_{0 \leq s \leq a_{T_{k,j}}} \frac{\|Y(t+s) - Y(t)\|_{l^p}}{\tilde{\sigma}(p, a_{T_{k,j}})\beta_1(k,j)}$$
$$\leq \theta^2 \quad \text{a.s.,}$$

where

$$\beta_1(k,j) = \left\{2\left(\log 2^j + \log\log \theta^{|k|}\right)\right\}^{1/2}.$$

Let $k \neq 0$. Then it follows from Lemma 2.4 that there exists $C_\varepsilon > 0$ such that

$$
\begin{aligned}
P\Big\{ &\sup_{j \geq l}\ \sup_{0 \leq t \leq b_{T_{k,j}}}\ \sup_{0 \leq s \leq a_{T_{k,j}}}\ \frac{\|Y(t+s) - Y(t)\|_{l^p}}{\tilde\sigma(p, a_{T_{k,j}})\beta_1(k,j)} \geq \theta \Big\} \\
&\leq C_\varepsilon \sum_{j \geq l} \left(\frac{b_{T_{k,j}}}{a_{T_{k,j}}}\right) \exp\left(C\,\theta\,\beta_1(k,j) - \frac{\theta^2}{2+\varepsilon}\,\beta_1^2(k,j)\right) \\
&\leq C_\varepsilon \sum_{j \geq l} (2^j)^{1-\frac{2\theta^2}{2+\varepsilon}}\ (\log\theta^{|k|})^{-\frac{2\theta^2}{2+\varepsilon}} \\
&\leq C_\varepsilon |k|^{-1-\frac{\varepsilon}{2+\varepsilon}}\ 2^{-\frac{\varepsilon}{2+\varepsilon}l}.
\end{aligned}
\tag{2.14}
$$

Hence we have

$$
\sum_{l=0}^{\infty}\sum_{|k|=1}^{\infty} P\Big\{\sup_{j \geq l}\ \sup_{0 \leq t \leq b_{T_{k,j}}}\ \sup_{0 \leq s \leq a_{T_{k,j}}}\ \frac{\|Y(t+s) - Y(t)\|_{l^p}}{\tilde\sigma(p, a_{T_{k,j}})\beta_1(k,j)} \geq \theta \Big\} < \infty.
\tag{2.15}
$$

If $k = 0$, then, by the same way as $(2.12)\sim(2.14)$, we have

$$
\begin{aligned}
P\Big\{ &\sup_{j \geq l}\ \sup_{0 \leq t \leq b_{T_{k,j}}}\ \sup_{0 \leq s \leq a_{T_{k,j}}}\ \frac{\|Y(t+s) - Y(t)\|_{l^p}}{\tilde\sigma(p, a_{T_{k,j}})\beta_1(0,j)} \geq \theta \Big\} \\
&\leq C_\varepsilon \sum_{j \geq l} \left(\frac{b_{T_{k,j}}}{a_{T_{k,j}}}\right)\ \exp\left(-\frac{\theta^2}{2+\varepsilon}\left(2\log 2^j\right)\right) \\
&\leq C_\varepsilon 2^{-\frac{\varepsilon}{2+\varepsilon}l},
\end{aligned}
\tag{2.16}
$$

where $\beta_1(0,j) = \{2\log 2^j\}^{1/2}$. Hence it follows that

$$
\sum_{l=0}^{\infty} P\Big\{\sup_{j \geq l}\ \sup_{0 \leq t \leq b_{T_{k,j}}}\ \sup_{0 \leq s \leq a_{T_{k,j}}}\ \frac{\|Y(t+s) - Y(t)\|_{l^p}}{\tilde\sigma(p, a_{T_{k,j}})\beta_1(0,j)} \geq \theta \Big\} < \infty.
\tag{2.17}
$$

Considering $(2.12)\sim(2.13)$ and applying the Borel-Cantelli lemma to (2.15) and (2.17), we obtain (1.5) by the arbitrariness of $\theta > 1$.

CASE 2: The case in which (2.12) holds when $j \leq -l < 0$.

By (i) and the definition of $A_{k,j}$, this case implies $|k| \to \infty$. We will prove that

$$
\limsup_{|k|+l\to\infty}\ \sup_{j \leq -l}\ \sup_{0 \leq t \leq b_{T_{k,j}}}\ \sup_{0 \leq s \leq a_{T_{k,j}}}\ \frac{\|Y(t+s) - Y(t)\|_{l^p}}{\tilde\sigma(p, a_{T_{k,j}})\beta_1(k,0)} \leq \theta \quad \text{a.s.,}
$$

where $\beta_1(k,0) = \{2\log\log\theta^{|k|}\}^{1/2}$. It follows from Lemma 2.4 that there exists $C_\varepsilon > 0$ such that

$$P\left\{\sup_{j\leq -l}\ \sup_{0\leq t\leq b_{T_{k,j}}}\ \sup_{0\leq s\leq a_{T_{k,j}}}\ \frac{\|Y(t+s)-Y(t)\|_{l^p}}{\tilde{\sigma}(p,a_{T_{k,j}})\beta_1(k,0)} > \theta\right\}$$

$$\leq C_\varepsilon \sum_{j\leq -l}\left(\frac{b_{T_{k,j}}}{a_{T_{k,j}}}\right)\exp\left(C\theta\beta_1(k) - \frac{\theta^2}{2+\varepsilon}\beta_1^2(k,0)\right)$$

$$\leq C_\varepsilon \sum_{j\leq -l} 2^j |k|^{-\frac{2\theta^2}{2+\varepsilon}} \leq C_\varepsilon 2^{-l}|k|^{-1-\frac{\varepsilon}{2+\varepsilon}}.$$

Hence we have

$$\sum_{l=1}^{\infty}\sum_{|k|=1}^{\infty} P\left\{\sup_{j\leq -l}\ \sup_{0\leq t\leq b_{T_{k,j}}}\ \sup_{0\leq s\leq a_{T_{k,j}}}\ \frac{\|Y(t+s)-Y(t)\|_{l^p}}{\tilde{\sigma}(p,a_{T_{k,j}})\beta_1(k,0)} > \theta\right\} < \infty.$$

As in CASE 1, we can get (1.5).

To prove Theorem 1.2, we need the following two lemmas:

Lemma 2.5. [8] *Let $\{\xi_j, j = 1, 2, \cdots, n\}$ be jointly standardized normal random variables with $\Lambda_{ij} = \mathrm{Corr}(\xi_i, \xi_j)$ such that $\delta := \max_{i\neq j}|\Lambda_{ij}| < 1$. Then, for any real number u and integers $1 \leq l_1 < l_2 < \cdots < l_k \leq n$ with $k \leq n$,*

$$(2.18)\qquad P\left\{\max_{1\leq i\leq k}\xi_{l_i} \leq u\right\} \leq \left(1-\Phi(u)\right)^k + K\sum_{1\leq i<j\leq k}|r_{ij}|\exp\left(\frac{-u^2}{1+|r_{ij}|}\right),$$

where $r_{ij} = \Lambda_{l_i l_j}$, K is a constant independent of n, u and k, and we denote

$$\Phi(t) = \int_t^{\infty}\frac{1}{\sqrt{2\pi}}e^{-y^2/2}\,dy.$$

Estimating an upper bound for the second term of the right hand side of (2.18), we can obtain the following lemma (cf. Lemma 4.4 in [2]):

Lemma 2.6. *Let $\{\xi_j, j = 1, 2, \cdots, n\}, \delta, k$ and r_{ij} be as in Lemma 2.5. Further assume that the inequality $|r_{ij}| < |i-j|^{-\nu}$ holds for some $\nu > 0$. Set $u = \{(2-\eta)\log k\}^{1/2}$, where $0 < \eta < (1-\delta)\nu/(1+\nu+\delta)$. Then we have*

$$\sum := \sum_{1\leq i<j\leq k}|r_{ij}|\exp\left(-\frac{u^2}{1+|r_{ij}|}\right) \leq ck^{-\delta_0},$$

where $\delta_0 = \{\nu(1-\delta) - \eta(1+\delta+\nu)\}/\{(1+\nu)(1+\delta)\} > 0$ and c is a positive constant independent of n, u and k.

The main stream of the proof of Theorem 1.2 is similar to that of Theorem B.

Proof of Theorem 1.2. Let $1 < \theta < 1 + \varepsilon$ for any given $0 < \varepsilon < 1$. Define

$$A_k = \{T : 2^k \le \frac{b_T}{a_T} \le 2^{k+1}\}, \quad k \ge 0,$$

$$A_{k,j} = \{T : \theta^{j-1} \le \tilde{\sigma}(p, a_T) \le \theta^j, \ T \in A_k\}, \quad -\infty < j < \infty,$$

$$b(T_{k,j}) = \inf\{b_T : T \in A_{k,j}\}, \quad a_{k,j} = \inf\{a_T : T \in A_{k,j}\}.$$

From (1.4) and (ii), we obtain

$$A_{k,j} = \phi \quad \text{for every} \ |j| \ge 2^{(2-\theta)(\theta-1)k}$$

provided that k is sufficiently large. Therefore, we have

$$\liminf_{T\to\infty} \sup_{0\le t\le b_T} \sup_{0\le s\le a_T} \frac{\|Y(t+s) - Y(t)\|_{l^p}}{\tilde{\sigma}(p, a_T)\beta_2(T)}$$

$$\ge \liminf_{k\to\infty} \min_{|j|<2^{(2-\theta)(\theta-1)k}} \sup_{0\le t\le b(T_{k,j})} \sup_{0\le s\le a_{k,j}} \frac{\|Y(t+s) - Y(t)\|_{l^p}}{\theta^j \sqrt{2\log 2^{k+1}}}$$

(2.19)
$$\ge \liminf_{k\to\infty} \min_{|j|<2^{(2-\theta)(\theta-1)k}} \max_{0\le i\le 2^{k(2-\theta)}}$$

$$\frac{\|Y(i2^{k(\theta-1)}a_{k,j} + a_{k,j}) - Y(i2^{k(\theta-1)}a_{k,j})\|_{l^p}}{\theta\tilde{\sigma}(p, a_{k,j})\sqrt{2\log 2^k}}$$

$$=: J.$$

We proceed with the proof by considering two cases of $1 \le p < 2$ and $2 \le p < \infty$, separately.

CASE I. $1 \le p < 2$. From (3.2) of [6], we have

(2.20)
$$\left(\sum_{n=1}^{\infty} |\xi_n|^p\right)^{1/p} = \sup_{\|a\|_{l^q}\le 1} \sum_{n=1}^{\infty} \xi_n a_n,$$

where $q = p/(p-1)$ and $a = (a_1, a_2, \cdots) \in l^q$. Thus we have

$$\|Y((i2^{k(\theta-1)} + 1)a_{k,j}) - Y(i2^{k(\theta-1)}a_{k,j})\|_{l^p}$$

(2.21)
$$\ge \left(\sum_{n=1}^{\infty} (\sigma_n(a_{k,j}))^{\frac{2p}{2-p}}\right)^{-(p-1)/p} \sum_{n=1}^{\infty} \{(\sigma_n(a_{k,j}))^{\frac{2(p-1)}{2-p}}$$

$$\times \left(X_n((i2^{k(\theta-1)} + 1)a_{k,j}) - X_n(i2^{k(\theta-1)}a_{k,j})\right)\}.$$

Consider

$$\xi(k,j;i)$$

$$(2.22) \quad := \frac{\sum_{n=1}^{\infty}\{\sigma_n(a_{k,j})^{\frac{2(p-1)}{2-p}}\big(X_n((i2^{k(\theta-1)}+1)a_{k,j}) - X_n(i2^{k(\theta-1)}a_{k,j})\big)\}}{\tilde{\sigma}(p,a_{k,j})\big(\sum_{n=1}^{\infty}(\sigma_n(a_{k,j}))^{\frac{2p}{2-p}}\big)^{(p-1)/p}}$$

$$= \frac{\sum_{n=1}^{\infty}\{\sigma_n(a_{k,j})^{\frac{2(p-1)}{2-p}}\big(X_n((i2^{k(\theta-1)}+1)a_{k,j}) - X_n(i2^{k(\theta-1)}a_{k,j})\big)\}}{\big(\sum_{n=1}^{\infty}(\sigma_n(a_{k,j}))^{\frac{2p}{2-p}}\big)^{1/2}}$$

for $k = 0,1,2,\cdots$, $|j| < 2^{(2-\theta)(\theta-1)k}$ and $0 \le i \le 2^{k(2-\theta)}$. Note that $\xi(k,j;i)$ are standard normal random variables. By (2.19), the inequality (1.6) follows if we show that

$$(2.23) \quad J \ge \frac{1}{\theta}\liminf_{k\to\infty}\min_{|j|<2^{(2-\theta)(\theta-1)k}}\max_{0\le i\le 2^{k(2-\theta)}}\frac{\xi(k,j;i)}{\sqrt{2\log 2^k}} \ge 1 \quad \text{a.s.}$$

For j,k fixed and $0 \le i < m \le 2^{k(2-\theta)}$, we have

$$|r(i,m)| := |Corr(\xi(k,j;i),\xi(k,j;m))|$$

$$= \left(\sum_{n=1}^{\infty}(\sigma_n(a_{k,j}))^{\frac{2p}{2-p}}\right)^{-1}\sum_{n=1}^{\infty}(\sigma_n(a_{k,j}))^{\frac{4(p-1)}{2-p}}$$

$$\times \Big|E\{\big(X_n(a_{k,j}) - X_n(0)\big)\big(X_n(((m-i)2^{k(\theta-1)}+1)a_{k,j})$$

$$- X_n((m-i)2^{k(\theta-1)}a_{k,j})\big)\}\Big|.$$

Using the elementary relation $2ab = (a^2+b^2-(a-b)^2)$ and (iii)$'$, it follows that

$$\Big|E\{\big(X_n(a_{k,j}) - X_n(0)\big)\big(X_n(((m-i)2^{k(\theta-1)}+1)a_{k,j})$$

$$- X_n((m-i)2^{k(\theta-1)}a_{k,j})\big)\}\Big|$$

$$= \Big| -\frac{1}{2}\{\sigma_n^2((m-i)2^{k(\theta-1)}a_{k,j}) - \sigma_n^2(((m-i)2^{k(\theta-1)}-1)a_{k,j})$$

$$- \sigma_n^2(((m-i)2^{k(\theta-1)}+1)a_{k,j}) + \sigma_n^2((m-i)2^{k(\theta-1)}a_{k,j})\}\Big|$$

$$\le \frac{1}{2}\int_{((m-i)2^{k(\theta-1)}-1)a_{k,j}}^{(m-i)2^{k(\theta-1)}a_{k,j}}\big|(\sigma_n^2(t+a_{k,j}))' - (\sigma_n^2(t))'\big|\,dt$$

$$\le \frac{1}{2}\int_{((m-i)2^{k(\theta-1)}-1)a_{k,j}}^{(m-i)2^{k(\theta-1)}a_{k,j}}\left(\int_{t}^{t+a_{k,j}}\big|(\sigma_n^2(y))''\big|\,dy\right)dt$$

$$\le \frac{1}{2}\int_{((m-i)2^{k(\theta-1)}-1)a_{k,j}}^{(m-i)2^{k(\theta-1)}a_{k,j}}\left(\int_{t}^{t+a_{k,j}} c\,\frac{\sigma_n^2(y)}{y^2}\,dy\right)dt$$

$$\le \frac{c}{2}\frac{\sigma_n^2(((m-i)2^{k(\theta-1)}+1)a_{k,j})}{((m-i)2^{k(\theta-1)}-1)^2}.$$

By the regularity of $\sigma_n(\cdot)$, we obtain $|r(i,m)| < |m - i|^{2\alpha - 2}$ provided that k is sufficiently large, where $\alpha = \sup\{\alpha_n : n \geq 1\}$. Let us now apply Lemmas 2.5 and 2.6 for

$$\xi_{l_i} = \xi(k,j;i), \quad |r_{ij}| = |r(i,m)| < |m - i|^{-\nu}, \quad \nu = 2 - \alpha, \quad 1 \leq i \leq 2^{k(2-\theta)},$$

$$u = \{(2 - \eta) \log 2^{k(2-\theta)}\}^{1/2}, \quad \eta = \frac{2 - 2\theta + 2\varepsilon}{2 - \theta} < \frac{(1 - \delta)\nu}{1 + \nu + \delta}.$$

Then we have

$$P\left\{ \min_{|j| < 2^{(2-\theta)(\theta-1)k}} \max_{0 \leq i \leq 2^{k(2-\theta)}} \frac{\xi(k,j;i)}{\sqrt{2 \log 2^k}} < \sqrt{1 - \varepsilon} \right\}$$

$$= P\left\{ \min_{|j| < 2^{(2-\theta)(\theta-1)k}} \max_{0 \leq i \leq 2^{k(2-\theta)}} \xi(k,j;i) < \sqrt{(2 - \eta)(2 - \theta) \log 2^k} \right\}$$

$$\leq \sum_{|j| < 2^{(2-\theta)(\theta-1)k}} P\left\{ \max_{0 \leq i \leq 2^{k(2-\theta)}} \xi(k,j;i) < \sqrt{(2 - \eta) \log 2^{k(2-\theta)}} \right\}$$

$$\leq \sum_{|j| < 2^{(2-\theta)(\theta-1)k}} \left\{ \left(1 - \Phi(u)\right)^{2^{k(2-\theta)}} + K 2^{-k(2-\theta)\delta_0} \right\}$$

$$\leq 2K 2^{(2-\theta)(\theta-1)k} \cdot 2^{-k(2-\theta)\delta_0} \leq 2K 2^{-\zeta k},$$

where $\zeta = (2 - \theta)(\delta_0 - \theta + 1) > 0$. The Borel-Cantelli lemma implies that (2.23) holds true since θ and ε are arbitrary.

CASE II. $2 \leq p < \infty$. Take $N_{k,j}$ such that $\sigma_{N_{k,j}}(a_{k,j}) = \sigma^*(a_{k,j})$. Clearly,

$$\frac{\|Y((i2^{k(\theta-1)} + 1)a_{k,j}) - Y(i2^{k(\theta-1)}a_{k,j})\|_{l^p}}{\tilde{\sigma}(p, a_{k,j})}$$

$$\geq \frac{X_{N_{k,j}}((i2^{k(\theta-1)} + 1)a_{k,j}) - X_{N_{k,j}}(i2^{k(\theta-1)}a_{k,j})}{\sigma_{N_{k,j}}(a_{k,j})}.$$

Consider

$$\zeta(k,j;i) = \frac{X_{N_{k,j}}((i2^{k(\theta-1)} + 1)a_{k,j}) - X_{N_{k,j}}(i2^{k(\theta-1)}a_{k,j})}{\sigma_{N_{k,j}}(a_{k,j})}$$

for $k = 0, 1, 2, \cdots$, $|j| < 2^{(2-\theta)(\theta-1)k}$ and $0 \leq i \leq 2^{k(2-\theta)}$. Then $\zeta(k,j;i)$ are standard normal random variables, and as in CASE I we can obtain

$$|R(i,m)| := |Corr(\zeta(k,j;i), \zeta(k,j;m))| < |m - i|^{2\alpha_{N_{k,j}} - 2}$$

provided that k is large enough. The rest of the proof is nearly the same as that of CASE I. Thus we conclude that (1.6) remains true in this case as well.

References

1. R. J. Adler, *An Introduction to Continuity, Extrema and Related Topics for General Gaussian Processes*, IMS, Hayward, CA, 1990.
2. Y. K. Choi, *Erdös-Rényi type laws applied to Gaussian processes*, J. Math. Kyoto Univ. **31** (1991), 191–217.
3. E. Csáki and M. Csörgő, *Inequalities for increments of stochastic processes and moduli of continuity*, Ann. Probab. **20** (1992), 1031–1052.
4. E. Csáki, M. Csörgő and Q. M. Shao, *Fernique type Inequalities and moduli of continuity for l^2-valued Ornstein-Uhlenbeck Processes*, Ann. Inst. Henri Poincare Probabilities et Statistiques **28** (1992), 479–517.
5. E. Csáki, M. Csörgő and Q. M. Shao, *Moduli of continuity for l^p-valued Gaussian processes*, Acta Sci. Math. **60** (1995), 149–175.
6. M. Csörgő and Q. M. Shao, *Strong limit theorems for large and small increments of l^p-valued Gaussian Processes*, Ann. Probab. **21** (1993), 1958–1990.
7. X. Fernique, *Continuité des processus Gaussiens*, C. R. Acad. Sci. Paris, t. **258** (1964), 6058–6060.
8. M. R. Leadbetter, G. Lindgren and H. Rootzen, *Extremes and Related Properties of Random Sequences and Processes*, Springer-Verlag, New York, 1983.
9. Z. Lin, *How big are the increments of l^p-valued Gaussian processes?* Science in China (Series A) **40(4)** (1997), 337–349.
10. Q. M. Shao, *p-variation of Gaussian processes with stationary increments*, Studia Sci. Math. Hungari. **31** (1996), 237–247.

LIMINF THEOREMS ON THE LAG
INCREMENTS OF A GAUSSIAN PROCESS

YONG KAB CHOI[1], SOON KYU PARK[1] AND KYO SHIN HWANG[2]

ABSTRACT. In this paper we establish some liminf theorems on the lag increments of a Gaussian process by using large deviation results of the Gaussian process.

1. Introduction and Results

Let $\{W(t), 0 \leq t < \infty\}$ be a standard Wiener process on a probability space $(\Omega, \mathcal{F}, P)$. We can find the following limsup theorem on the Wiener process from Csörgő and Révész [4].

Theorem A. *Let $0 < a_T \leq T$ be a function of T for which*

  (i) a_T *is nondecreasing,*

  (ii) T/a_T *is nondecreasing.*

Then

$$\limsup_{T \to \infty} \frac{|W(T + a_T) - W(T)|}{\sqrt{a_T}\, \beta_1(T)} = 1 \quad \text{a.s.,}$$

$$\limsup_{T \to \infty} \sup_{0 \leq t \leq T - a_T} \sup_{0 \leq s \leq a_T} \frac{|W(t + s) - W(t)|}{\sqrt{a_T}\, \beta_1(T)} = 1 \quad \text{a.s.,}$$

where $\beta_1(T) = \left\{2(\log(T/a_T) + \log\log T)\right\}^{1/2}$.

Here, and in the sequel, we shall define $\log t = \ln(\max\{t, 1\})$.

Limsup results on another form of increments, lag increments of a standard Wiener process were initially presented and discussed by Hanson and Russo [5]. Chen, Kong and Lin [1] sharpened their results and proved

Received November 9, 2000.

1991 *Mathematics Subject Classification.* 60F15, 60G15.

Key Words and Phrases. Lag increments, Wiener process, Gaussian process.

This research was supported by Chinju National University, 2000.

[1] Department of Mathematics, Geongsang National University, Chinju 660-701 Korea. Email addresses: mathykc@nongae.gsnu.ac.kr and kshwang@gshp.gsnu.ac.kr

[2] Department of Computer Science & Engineering, Chinju National University, 150, Chilam-Dong, Chinju, Kyungnam, Korea. Email: sgpark@chinju.ac.kr and sgpark@jinju.ac.kr

Theorem B. *For a_T such that $0 < a_T \leq T$, we have*

$$\limsup_{T \to \infty} \sup_{a_T \leq t \leq T} \frac{|W(T) - W(T-t)|}{D(T,t)} = 1 \quad \text{a.s.,}$$

$$\limsup_{T \to \infty} \sup_{a_T \leq t \leq T} \sup_{t \leq s \leq T} \frac{|W(s) - W(s-t)|}{D(T,t)} = 1 \quad \text{a.s.,}$$

$$\limsup_{T \to \infty} \sup_{a_T \leq t \leq T} \sup_{t \leq s \leq T} \sup_{0 \leq h \leq t} \frac{|W(s) - W(s-h)|}{D(T,t)} = 1 \quad \text{a.s.,}$$

where $D(T,t) = \left\{ 2t\left(\log(T/t) + \log\log t \right) \right\}^{1/2}$.

On the other hand, Shao [7] obtained the following liminf result of Theorem A.

Theorem C. *Let a_T be as in Theorem A. If*

$$\text{(iii)} \qquad\qquad \lim_{T \to \infty} \frac{T/a_T}{\log\log T} = \infty,$$

then

$$\liminf_{T \to \infty} \sup_{0 \leq t \leq T - a_T} \frac{|W(t + a_T) - W(t)|}{\sqrt{a_T}\, \gamma_T} = 1 \quad \text{a.s.,}$$

$$\liminf_{T \to \infty} \sup_{0 \leq t \leq T - a_T} \sup_{0 \leq s \leq a_T} \frac{|W(t + s) - W(t)|}{\sqrt{a_T}\, \gamma_T} = 1 \quad \text{a.s.,}$$

where $\gamma_T = \left\{ 2\left(\log(T/a_T) - \log\log\log T \right) \right\}^{1/2}$.

Also, He and Chen [6] investigated the inferior limit version of Theorem B.

Theorem D. *Let $0 < a_T \leq T$ be a nondecreasing function of T and satisfy*

$$\lim_{T \to \infty} \frac{\log(T/a_T)}{\log\log T} = r, \quad 0 \leq r \leq \infty.$$

Then

$$\liminf_{T \to \infty} \sup_{a_T \leq t \leq T} \sup_{t \leq s \leq T} \frac{|W(s) - W(s-t)|}{D(T,t)} = \beta_r \quad \text{a.s.,}$$

$$\liminf_{T \to \infty} \sup_{a_T \leq t \leq T} \sup_{t \leq s \leq T} \sup_{0 \leq h \leq t} \frac{|W(s) - W(s-h)|}{D(T,t)} = \beta_r \quad \text{a.s.,}$$

where $\beta_r = \{r/(1+r)\}^{1/2}$.

Let $\{X(t), 0 \le t < \infty\}$ be an almost surely continuous, centered Gaussian process on the probability space $(\Omega, \mathcal{F}, P)$ with $X(0) = 0$ and $\sigma^2(|t - s|) := E\{X(t) - X(s)\}^2$, where $\sigma(\cdot)$ is a nondecreasing continuous and regularly varying function on $(0, \infty)$ with exponent α at ∞ for some $0 < \alpha < 1$. Denote

$$d(T, t) = \{2\sigma^2(t)\big(\log(T/t) + \log\log t\big)\}^{1/2}.$$

We note that a Wiener process is a Gaussian process, but the converse is not true in general.

Choi and Hwang [3] generalized Theorems A and B, separately, as follows.

Theorem 1.1. *Let a_T be as in Theorem A. Assume that, for $t > 0$, either*
  (i) $\sigma^2(t)$ is concave
or
  (ii) $\sigma(t)$ is differentiable which satisfies

$$0 < \sigma'(t) \le c_1\sigma(t)/t, \quad |\sigma''(t)| \le c_2\sigma(t)/t^2, \quad \sigma'''(t) \le c_3\sigma(t)/t^3$$

for positive constants c_1, c_2 and c_3. Then

$$\limsup_{T\to\infty} \frac{|X(T) - X(T - a_T)|}{\sigma(a_T)\,\beta_1(T)} = 1 \quad a.s.,$$

$$\limsup_{T\to\infty} \sup_{0\le t\le T} \sup_{0\le s\le a_T} \frac{|X(t+s) - X(t)|}{\sigma(a_T)\,\beta_1(T)} = 1 \quad a.s.$$

Theorem 1.2. *Let a_T and $\sigma(t)$ be as in Theorem 1.1. Then*

$$\limsup_{T\to\infty} \frac{|X(T) - X(T - a_T)|}{d(T, a_T)} = 1 \quad a.s.,$$

$$\limsup_{T\to\infty} \sup_{a_T\le t\le T} \sup_{t\le s\le T} \frac{|X(s) - X(s - t)|}{d(T, t)} = 1 \quad a.s.,$$

$$\limsup_{T\to\infty} \sup_{a_T\le t\le T} \sup_{t\le s\le T} \sup_{0\le h\le t} \frac{|X(s) - X(s - h)|}{d(T, t)} = 1 \quad a.s.$$

The main objective of this paper is to extend Theorems C and D to the following Theorems 1.3 and 1.4 on the Gaussian process, respectively. Our results are as follows:

Theorem 1.3. *Let a_T $(0 < T < \infty)$ be a nondecreasing function of T such that*

(i) $$0 < a_T \leq T,$$

(ii) $$\lim_{T \to \infty} \frac{\log(T/a_T)}{\log_{(m+1)} T} = \infty, \quad m \geq 1.$$

Suppose that there exists a positive constant c_1 such that

(iii) $$\left| \frac{d^2 \sigma^2(x)}{dx^2} \right| \leq c_1 \frac{\sigma^2(x)}{x^2}, \quad x > 0.$$

Then we have, for $i = 1, 2, 3$

(1.1) $$\liminf_{T \to \infty} \sup_{0 \leq t \leq T - a_T} \frac{|X(t + a_T) - X(t)|}{\sigma(a_T)\,\beta_i(T)} = 1 \quad \text{a.s.},$$

(1.2) $$\liminf_{T \to \infty} \sup_{0 \leq t \leq T} \sup_{0 \leq s \leq a_T} \frac{|X(t + s) - X(t)|}{\sigma(a_T)\,\beta_i(T)} = 1 \quad \text{a.s.},$$

where

$$\beta_2(T) = \left\{ 2\big(\log(T/a_T) + \log_{(m+1)} T \big) \right\}^{1/2},$$

$$\beta_3(T) = \left\{ 2\big(\log(T/a_T) - \log_{(m+2)} T \big) \right\}^{1/2}, \quad m \geq 1.$$

Note that $\beta_3(T) \leq \beta_2(T) \leq \beta_1(T)$ for all large $T > 0$.

Theorem 1.4. *Let a_T $(0 < T < \infty)$ be a nondecreasing function of T such that*

(i) $$0 < a_T \leq T,$$

(ii)′ $$\lim_{T \to \infty} \frac{\log(T/a_T)}{\log \log T} = r, \quad 0 \leq r \leq \infty.$$

Assume that $\sigma(t)$, $t > 0$, is differentiable which satisfies

(iii) $$0 < \sigma'(t) \leq c_1 \sigma(t)/t, \quad |\sigma''(t)| \leq c_2 \sigma(t)/t^2, \quad \sigma'''(t) \leq c_3 \sigma(t)/t^3$$

for positive constants c_1, c_2 and c_3. Then we have

(1.3) $$\liminf_{T \to \infty} \sup_{a_T \leq t \leq T} \sup_{t \leq s \leq T} \frac{|X(s) - X(s - t)|}{d(T, t)} = \beta_r \quad \text{a.s.},$$

(1.4) $$\liminf_{T \to \infty} \sup_{a_T \leq t \leq T} \sup_{t \leq s \leq T} \sup_{0 \leq h \leq t} \frac{|X(s) - X(s - h)|}{d(T, t)} = \beta_r \quad \text{a.s.},$$

where $\beta_r = \{ r/(1 + r) \}^{1/2}$.

2. The Proofs

The following Lemma 2.1 is obvious.

Lemma 2.1. *Let $\{\xi, \xi_n : n \geq 1\}$ be a sequence of random variables. If*

$$P\{\xi_n > \xi\} \longrightarrow 0 \quad \text{as } n \to \infty,$$

then there exists a subsequence $\{\xi_{n_k}\}_{k=1}^{\infty}$ such that

$$\limsup_{k \to \infty} \xi_{n_k} \leq \xi \quad \text{a.s.}$$

So we have

$$\liminf_{n \to \infty} \xi_n \leq \xi \quad \text{a.s.}$$

Lemma 2.2. [2] *For any $\epsilon > 0$ there exists a constant $C_\epsilon > 0$ depending only on ϵ such that, for all $u \geq 0$,*

$$P\left\{ \sup_{0 \leq t \leq T} \sup_{0 \leq s \leq a_T} \frac{|X(t+s) - X(t)|}{\sigma(a_T)} > u \right\} \leq C_\epsilon \frac{T}{a_T} e^{-u^2/(2+\epsilon)}.$$

Proof of Theorem 1.3. The condition (ii) implies that for any $\lambda > 0$ we have

$$\frac{T}{a_T} > \left(\log_{(m)} T \right)^{\lambda}$$

for all large $T > 0$. Thus it follows from Lemma 2.2 that for any $\epsilon > 0$

$$P\left\{ \sup_{0 \leq t \leq T} \sup_{0 \leq s \leq a_T} \frac{|X(t+s) - X(t)|}{\sigma(a_T)\,\beta_3(T)} > \sqrt{1+2\epsilon} \right\}$$

$$\leq C_\epsilon \frac{T}{a_T} \exp\left(-\frac{2+4\epsilon}{2+\epsilon} \log\left(\frac{T}{a_T}\left(\log_{(m+1)} T\right)^{-1} \right) \right)$$

$$\leq C_\epsilon \left(\log_{(m)} T \right)^{-\lambda \epsilon'} \left(\log_{(m+1)} T \right)^{1+\epsilon'}, \quad \epsilon' = \frac{3\epsilon}{2+\epsilon} > 0$$

$$\longrightarrow 0 \quad \text{as } T \to \infty.$$

Applying Lemma 2.1, we have

$$(2.1) \qquad \liminf_{T \to \infty} \sup_{0 < t \leq T} \sup_{0 \leq s \leq a_T} \frac{|X(t+s) - X(t)|}{\sigma(a_T)\,\beta_3(T)} \leq 1 \quad \text{a.s.}$$

On the other hand, the lower bound

$$(2.2) \qquad \liminf_{T \to \infty} \sup_{0 < t \leq T - a_T} \frac{|X(t+a_T) - X(t)|}{\sigma(a_T)\,\beta_1(T)} \geq 1 \quad a.s.$$

follows immediately from Theorem 2.2 of Choi [2]. Therefore (1.1) and (1.2) hold true from (2.1) and (2.2).

Proof of Theorem 1.4. The main stream of the proof is similar to the proof of Theorem D. First we prove that

$$(2.3) \qquad \liminf_{T\to\infty} \; \sup_{a_T\le t\le T} \; \sup_{t\le s\le T} \frac{|X(s)-X(s-t)|}{d(T,t)} \ge \beta_r \quad \text{a.s.}$$

Clearly, (2.3) is true for $r = 0$. In case of $0 < r \le \infty$, (ii)$'$ implies (ii) when $m \ge 2$. Thus it follows from (1.1) that, for $m \ge 2$,

$$\liminf_{T\to\infty} \; \sup_{a_T\le t\le T} \; \sup_{t\le s\le T} \frac{|X(s)-X(s-t)|}{d(T,t)}$$

$$\ge \liminf_{T\to\infty} \; \sup_{0\le s\le T-a_T} \frac{|X(s+a_T)-X(s)|}{d(T,a_T)}$$

$$\ge \liminf_{T\to\infty} \left(\frac{2\sigma^2(a_T)\big(\log(T/a_T) + \log_{(m+1)} T\big)}{2\sigma^2(a_T)\big(\log(T/a_T) + \log\log T\big)} \right)^{1/2}$$

$$= \beta_r \quad \text{a.s.}$$

Now, (1.3) and (1.4) follow from (2.3) if we show that

$$(2.4) \qquad \liminf_{T\to\infty} \; \sup_{a_T\le t\le T} \; \sup_{t\le s\le T} \; \sup_{0\le h\le t} \frac{|X(s)-X(s-h)|}{d(T,t)} \le \beta_r \quad \text{a.s.}$$

Clearly, (2.4) holds for $r = \infty$ from Theorem 1.2. We need only to prove that

$$(2.5) \qquad \liminf_{n\to\infty} \; \sup_{a_{T_n}\le t\le T_n} \; \sup_{t\le s\le T_n} \; \sup_{0\le h\le t} \frac{|X(s)-X(s-h)|}{d(T_n,t)} \le \beta_r \quad \text{a.s.}$$

for $T_n = e^{e^n}$ and $0 \le r < \infty$. For any $\beta_r < \beta < 1$, we take $\theta > 1$, $0 < \alpha < 1$ and $\epsilon > 0$ such that

$$\frac{2\beta^2}{(2+\epsilon)\theta^{2\alpha}} > \frac{r}{1+r} + \epsilon.$$

Put $K_n = \big[\log_\theta \frac{T_n}{a_{T_n}}\big]$, $t_k = \theta^k a_{T_n}$, $0 \le k \le K_n$. We have

$$\sup_{a_{T_n}\le t\le T_n} \; \sup_{t\le s\le T_n} \; \sup_{0\le h\le t} \frac{|X(s)-X(s-h)|}{d(T_n,t)}$$

$$\le \max_{0\le k\le K_n} \; \sup_{t_k\le t\le t_{k+1}} \; \sup_{t\le s\le T_n} \; \sup_{0\le h\le t} \frac{|X(s)-X(s-h)|}{d(T_n,t_k)}$$

$$\le \max_{0\le k\le K_n} \; \sup_{0\le s-h, \; s\le T_n} \; \sup_{0\le h\le t_{k+1}} \frac{|X(s)-X(s-h)|}{d(T_n,t_k)}$$

$$=: \max_{0\le k\le K_n} A_{nk}.$$

It follows from the condition (ii)$'$ and Lemma 2.2 that, for large n

$$P\{A_{nk} \geq \beta\} \leq C\frac{T_n}{t_{k+1}} \exp\left(-\frac{2\beta^2}{2+\epsilon} \cdot \frac{\sigma^2(t_k)}{\sigma^2(t_{k+1})}\left(\log\frac{T_n}{t_k} + \log\log t_k\right)\right)$$

$$\leq C\left(\log T_n\right)^{r+\epsilon-(r+\epsilon+1)2\beta^2/((2+\epsilon)\theta^{2\alpha})}\theta^{-k(1-\beta^2)}$$

$$\leq C\,e^{-n\epsilon^2}\theta^{-k(1-\beta^2)}.$$

Therefore

$$\sum_{n=1}^{\infty} P\left\{\max_{0\leq k\leq K_n} A_{nk} \geq \beta\right\} < \infty$$

and the Borel-Cantelli lemma yields (2.5).

REFERENCES

1. G. J. Chen, F. C. Kong and Z. Y. Lin, *Answers to some questions about increments of a Wiener process*, Ann. Probab. **14** (1986), 1252–1261.
2. Y. K. Choi, *Erdös-Rényi type laws applied to Gaussian processes*, J. Math. Kyoto Univ. **31-1** (1991), 191–217.
3. Y. K. Choi and K. S. Hwang, *How big are the lag increments of a Gaussian process?*, Comp. & Math. Appl. **40** (2000), 911-919.
4. M. Csörgő and P. Révész, *How big are the increments of a Wiener process?*, Ann Probab. **7** (1979), 731-737.
5. D. L. Hanson and R. P. Russo, *Some results on increments of the Wiener process with applications to lag sums of i.i.d. r.v.s.*, Ann. Probab. **11** (1983), 609–623.
6. F. X. He and B. Chen, Some results on increments of the Wiener process, Chinese J. Appl. Probab. Statist. **5** (1989), 317–326 (in Chinese).
7. Q. M. Shao, *Remark on increments of the Wiener process*, J. Math. **6** (1986), 175–182 (in Chinese).

LIMINF RESULTS ON THE INCREMENTS OF A d-DIMENSIONAL GAUSSIAN PROCESS WITH INDEPENDENT COPIES

KYO SHIN HWANG[1]

ABSTRACT. In this paper we establish limit inferior results on the increments of a d-dimensional vector-valued Gaussian process with independent copies, via estimating upper bounds of large deviation probabilities on the suprema of the d-dimensional vector-valued Gaussian process.

1. Introduction and Results

The liminf theorems on the increments of Wiener processes or Gaussian processes have been studied in various ways by many authors, for instance, Book and Shore [1], Csáki and Révész [4], Csörgő and Révész [5], Deo [6], Hong [8], Shao [12], Zhang [14] and etc. In this paper we establish limit inferior results on the increments of a d-dimensional vector-valued Gaussian process with independent copies under mild conditions, via estimating upper bounds of large deviation probabilities on the suprema of the d-dimensional Gaussian process.

For the Wiener process, Csáki and Révész [4] obtained the following thoerem:

Theorem A. *Let* $\{W(t) : t \geq 0\}$ *be a standard Wiener process. Let* a_T $(0 < T < \infty)$ *be a function of* T *for which*
(a) $0 < a_T \leq T$,
(b) a_T *is non-decreasing,*
(c) T/a_T *is non-decreasing,*
(d) $\lim_{T \to \infty} \dfrac{\log (T/a_T)}{\log \log \log T} = \infty.$

Received November 9, 2000.

1991 *Mathematics Subject Classification.* Primary 60F15; Secondary 60G15.

Key Words and Phrases. Gaussian process, Wiener process, fractional Brownian motion, regular varying function.

This work was supported by Korea Research Foundation Grant (KRF-99-005-D00003).

[1] Department of Mathematics, Colleage of Natural Science, Geongsang National University, Chinju 660-701 Korea. Email: hwang0412@hanmail.net

Then

$$\liminf_{T\to\infty} \gamma_1(T) \sup_{0\le t\le T-a_T} \sup_{0\le s\le a_T} |W(t+s)-W(t)| = 1 \quad \text{a.s.,}$$

where

$$\gamma_1(T) = \left\{2a_T \log\left(1 + \frac{\pi^2}{16}\frac{T}{a_T \log\log T}\right)\right\}^{-1/2}.$$

For the fractional Brownian motion, Zhang [14] proved the following:

Theorem B. *Let $\{X(t) : t \ge 0\}$ be a fractional Brownian motion order 2α with $0 < \alpha < 1$. Let a_T be a function of T satisfying (a), (b), (c) and*

(e) $\lim_{T\to\infty} \dfrac{T/a_T}{\log\log T} = \infty.$

Then

$$\liminf_{T\to\infty} \gamma_2(T) \sup_{0\le t\le T-a_T} \sup_{0\le s\le a_T} |X(t+s)-X(t)| = 1 \quad \text{a.s.,}$$

where

$$\gamma_2(T) = a_T^{-\alpha}\left\{2\big(\log(T/a_T) - \log\log\log T\big)\right\}^{-1/2}.$$

In this paper, we shall establish liminf results on the increments of a d-dimensional Gaussian process under weaker conditions than (e). Throughout the paper we shall always assume the following statements: Let $\{X(t), t \in [0,\infty)\}$ be a real-valued continuous centered Gaussian process with $X(0) = 0$ and $E\{X(t) - X(s)\}^2 = \sigma^2(|t - s|)$, where $\sigma(t), t > 0$, is a positive nondecreasing continuous and regularly varying function with exponent α' at 0 and exponent α'' at ∞ ($0 < \alpha', \alpha'' < 1$). And let $\{X^d(t) = (X_1(t),\cdots,X_d(t)), t \ge 0\}$ be a d-dimensional Gaussian process whose components $X_i(t)$, $i = 1, 2,\cdots, d$, are copies of $X(t)$.

For $0 < T < \infty$, let a_T and b_T be real-valued positive functions of T such that

(i) $\qquad\qquad\qquad\qquad 0 < a_T \le T, \quad \liminf_{T\to\infty} a_T > 0.$

For convenience, we denote:

$$\beta_1(T) = \left\{2\left(\log(T/a_T) + \log\left(b_T + \frac{1}{b_T}\right)\right)\right\}^{1/2},$$

$$\beta_2(T) = \left\{2\big(\log(T/a_T) - \log_{(m+2)} T\big)\right\}^{1/2}, \quad m \ge 2,$$

where $\log_{(m)} T = \log\log_{(m-1)} T$, $m \ge 2$, $\log_{(1)} T = \log T$, $\log x = \ln(\max\{x, 1\})$.

The main results are as follows:

Theorem 1.1. *Suppose that*

(ii)
$$\lim_{T \to \infty} \left(\frac{T}{a_T} + b_T + \frac{1}{b_T} \right) = \infty.$$

Then we have

(1.1)
$$\liminf_{T \to \infty} \sup_{0 \le t \le T} \sup_{0 \le s \le a_T} \frac{\|X^d(t+s) - X^d(t)\|}{\sigma(a_T)\,\beta_1(T)} \le 1 \quad a.s.$$

Theorem 1.2. *Suppose that*

(iii)
$$\lim_{T \to \infty} \frac{T/a_T}{\log_{(m)} T} = \infty, \quad m \ge 2.$$

Then we have

(1.2)
$$\liminf_{T \to \infty} \sup_{0 \le t \le T} \sup_{0 \le s \le a_T} \frac{\|X^d(t+s) - X^d(t)\|}{\sigma(a_T)\,\beta_2(T)} \le 1 \quad a.s.$$

Theorem 1.3. *Assume that $X_i(t)$, $i = 1, 2, \cdots, d$, are independent copies of $X(t)$. Suppose that*

(iv)
$$\lim_{T \to \infty} \frac{\log(T/a_T)}{\log\left(b_T + \frac{1}{b_T}\right)} = \infty,$$

and there exist positive constants c_1 and c_2 such that, for $x > 0$,

(v)
$$\left| \frac{d\sigma^2(x)}{dx} \right| \le c_1 \frac{\sigma^2(x)}{x} \quad \text{and} \quad \left| \frac{d^2\sigma^2(x)}{dx^2} \right| \le c_2 \frac{\sigma^2(x)}{x^2}.$$

Then we have

(1.3)
$$\liminf_{T \to \infty} \sup_{0 \le t \le T} \sup_{0 \le s \le a_T} \frac{\|X^d(t+s) - X^d(t)\|}{\sigma(a_T)\,\beta_1(T)} \ge 1 \quad a.s.$$

From Theorems 1.1~1.3, we get the following liminf results:

Corollary 1.1. *For $m \ge 2$, let $b_T + \frac{1}{b_T} \ge \log_{(m)} T$ in (iv). Under the assumptions of Theorem 1.3, we have*

$$\liminf_{T \to \infty} \sup_{0 \le t \le T} \sup_{0 \le s \le a_T} \frac{\|X^d(t+s) - X^d(t)\|}{\sigma(a_T)\,\beta_j(T)} = 1, \quad j = 1, 2, \quad a.s.$$

Example 1. Let $a_T = T/\log T$ and $b_T = \log\log T$. Then we have

$$\liminf_{T\to\infty} \sup_{0\le t\le T} \sup_{0\le s\le T/\log T} \frac{\|X^d(t+s) - X^d(t)\|}{\sigma(T/\log T)\sqrt{\log\log T}} = \sqrt{2} \quad \text{a.s.}$$

Corollary 1.2. *Under the assumptions of Theorem 1.3,*

$$\liminf_{T\to\infty} \sup_{0\le t\le T} \sup_{0\le s\le a_T} \frac{\|X^d(t+s) - X^d(t)\|}{\sigma(a_T)\,\beta_1(T)} = 1 \quad \text{a.s.}$$

Example 2. Let $a_T = b_T = 1$. Then we have

$$\liminf_{T\to\infty} \sup_{0\le t\le T} \sup_{0\le s\le 1} \frac{\|X^d(t+s) - X^d(t)\|}{\sqrt{\log T}} = \sqrt{2}\sigma(1) \quad \text{a.s.}$$

2. The Proofs

Let $\mathbb{D}$ be a compact subset of $[0,\infty)^N$ with the Euclidean norm $\|\cdot\|$, and let $\{X(\mathbf{t}), \mathbf{t} \in \mathbb{D}\}$ be a real-valued separable, centered Gaussian process with $E\{X(\mathbf{t}) - X(\mathbf{s})\}^2 = \sigma^2(\|\mathbf{t} - \mathbf{s}\|)$. Suppose that

$$(2.1) \qquad 0 < \Gamma := \sup_{\mathbf{t}\in\mathbb{D}} \left\{E(X(\mathbf{t}))^2\right\}^{1/2} < \infty$$

and

$$(2.2) \qquad \sigma(h) \le \varphi(h),$$

where $\varphi(\cdot)$ is a nondecreasing continuous function. Let $\{X^d(\mathbf{t}) = (X_1(\mathbf{t}), \cdots, X_d(\mathbf{t})), \mathbf{t} \in \mathbb{D}\}$ be a d-dimensional Gaussian process whose components $X_i(\mathbf{t})$, $i = 1, 2, \cdots, d$, are copies of $X(\mathbf{t})$. Denote the Lebesgue measure of $\mathbb{D}$ by $m(\mathbb{D})$.

The following lemma is a Fernigue type inequality [7] on the d-dimensional Gaussian process, whose proof is similar to that of Lemma 2.3 in [3]:

Lemma 2.1. *For $\lambda > 0$, $x \ge 1$ and $\mathcal{B} > (2\sqrt{2} + 2)\sqrt{2N\log 2}$, we have*

$$P\left\{\sup_{\mathbf{t}\in\mathbb{D}} \|X^d(\mathbf{t})\| > x\left(\Gamma + \mathcal{B}\int_0^\infty \varphi(\sqrt{N}\lambda 2^{-y^2})\,dy\right)\right\} \le C\frac{m(\mathbb{D})}{\lambda^N}\Phi_d(x),$$

where $C > 0$ is a constant, $\Phi_d(x) = P\{\|N^d(0,1)\| \ge x\}$ and $N^d(0,1)$ denotes a d-dimensional standardized normal random vector.

Using Lemma 2.1, we can estimate an upper bound of the following large deviation probabilities by the same method as the proof of Lemma 2.4 in [3].

Lemma 2.2. *Let a_T be a positive function of $T > 0$. Then, for any $\varepsilon > 0$, there exists a positive constant C_ε depending only on ε such that*

$$P\left\{ \sup_{0 \le t \le T} \sup_{0 \le s \le a_T} \frac{\|X^d(t+s) - X^d(t)\|}{\sigma(a_T)} \ge x \right\}$$
$$\le C_\varepsilon \left(\frac{T}{a_T} \right) \Phi_d \left(\frac{2}{2+\varepsilon} x \right)$$

for all $x > 0$.

Lemma 2.3. ([11]) *Let $\{\xi, \xi_n; n \ge 1\}$ be a sequence of random variables. If*

$$P\{\xi_n \ge \xi\} \longrightarrow 0 \quad \text{as } n \to \infty,$$

then there is a subsequence $\{\xi_{n_k}\}$ such that

$$\limsup_{k \to \infty} \xi_{n_k} \le \xi \quad \text{a.s.}$$

So we have

$$\liminf_{n \to \infty} \xi_n \le \xi \quad \text{a.s.}$$

Proof of Theorem 1.1. By virtue of (ii), let $\{T_n\}_{n=1}^\infty$ be a sequence such that $T_n \le T \le T_{n+1}$ and

$$(2.7) \qquad \frac{T_n}{a_{T_n}} \longrightarrow \infty \quad \text{or} \quad b_{T_n} + \frac{1}{b_{T_n}} \longrightarrow \infty \quad \text{as } n \to \infty.$$

Note that, for sufficiently large $x > 0$, we have

$$(2.8) \qquad \Phi_d(x) \le C x^{d-2} e^{-x^2/2} \le C \exp\left(-\frac{x^2}{2+\varepsilon} \right)$$

for some $C > 0$ (see Lemma 1 in [9]). By (2.7), (2.8) and Lemma 2.2, we have, for $0 < \varepsilon' < \varepsilon < 1$,

$$P\left\{ \sup_{0 \le t \le T_n} \sup_{0 \le s \le a_{T_n}} \frac{\|X^d(t+s) - X^d(t)\|}{\sigma(a_{T_n})\beta_1(T_n)} \ge \sqrt{1+2\varepsilon} \right\}$$
$$\le C_\varepsilon \left(\frac{T_n}{a_{T_n}} \right) \Phi_d \left(\frac{2}{2+\varepsilon} \left[2(1+2\varepsilon) \log \left(\frac{T_n}{a_{T_n}} \left(b_{T_n} + \frac{1}{b_{T_n}} \right) \right) \right]^{1/2} \right)$$
$$\le C_\varepsilon \left(\frac{T_n}{a_{T_n}} \right) \exp\left(-(1+\varepsilon') \log \left(\frac{T_n}{a_{T_n}} \left(b_{T_n} + \frac{1}{b_{T_n}} \right) \right) \right)$$
$$= C_\varepsilon \left(\frac{T_n}{a_{T_n}} \right)^{-\varepsilon'} \left(b_{T_n} + \frac{1}{b_{T_n}} \right)^{-1-\varepsilon'} \longrightarrow 0 \quad \text{as } n \to \infty$$

and, further, by Lemma 2.3,

$$\liminf_{n\to\infty} \sup_{0\le t\le T_n} \sup_{0\le s\le a_{T_n}} \frac{\|X^d(t+s) - X^d(t)\|}{\sigma(a_{T_n})\,\beta_1(T_n)} \le \sqrt{1+\varepsilon} \quad \text{a.s.}$$

Since $\varepsilon > 0$ is arbitrary, we obtain (1.1).

Proof of Theorem 1.2. Considering (iii), we can take a sequence $\{T_n\}_{n=1}^{\infty}$ such that $T_n \le T \le T_{n+1}$ and, for any $\tau > 0$, there exists n_0 such that

$$(T_n/a_{T_n})\big/\log_{(m)} T_n > \tau \qquad \text{for all} \quad n \ge n_0.$$

Again by (2.8) and Lemma 2.2 we have, for $0 < \varepsilon' < \varepsilon < 1$,

$$\begin{aligned}
P&\left\{ \sup_{0\le t\le T_n} \sup_{0\le s\le a_{T_n}} \frac{\|X^d(t+s) - X^d(t)\|}{\sigma(a_{T_n})\beta_2(T_n)} \ge \sqrt{1+2\varepsilon} \right\} \\
&\le C_\varepsilon\left(\frac{T_n}{a_{T_n}}\right) \exp\left(-(1+\varepsilon')\log\left(\frac{T_n}{a_{T_n}}\frac{1}{\log_{(m+1)} T_n}\right)\right) \\
&= C_\varepsilon\left(\frac{T_n}{a_{T_n}}\right)^{-\varepsilon'} \left(\log_{(m+1)} T_n\right)^{1+\varepsilon'} \\
&\le C_\varepsilon\left(\tau\log_{(m)} T_n\right)^{-\varepsilon'} \left(\log_{(m+1)} T_n\right)^{1+\varepsilon'} \\
&\longrightarrow 0 \quad \text{as} \quad n\to\infty.
\end{aligned}$$

Thus Lemma 2.3 implies that (1.2) holds true.

The following Lemmas 2.4~2.7 are needed to prove Theorem 1.3. The proof of Lemma 2.4 is similar to that of Lemma 5 in [2].

Lemma 2.4. *Let $\sigma^2(x)$ be a nondecreasing continuous function of $x > 0$. Assume that there exist positive constants c_1 and c_2 such that*

$$\left|\frac{d\sigma^2(x)}{dx}\right| \le c_1\frac{\sigma^2(x)}{x}, \qquad \left|\frac{d^2\sigma^2(x)}{dx^2}\right| \le c_2\frac{\sigma^2(x)}{x^2}.$$

Let m be a positive number and a, b be real numbers. Then there exists a constant $c > 0$ such that

$$\left| \int_{|a(mb)|}^{|a(mb+1)|} d\sigma^2(x) - \int_{|a(mb-1)|}^{|a(mb)|} d\sigma^2(x) \right| \le c\frac{\sigma^2(|a(mb+1)|)}{|(mb-1)|^2}.$$

The following lemma is Corollary 4.2.4 in [10]:

Lemma 2.5. *Let $\{\xi_j, j = 1, 2, \cdots, n\}$ be jointly standardized normal random variables with $\Lambda_{ij} := \mathrm{Corr}\,(\xi_i, \xi_j)$ such that $\delta := \max_{i \neq j} |\Lambda_{ij}| < 1$. Then for any real number u and integers $1 \leq l_1 < l_2 < \cdots < l_k \leq n$ with $k \leq n$,*

$$(2.9) \qquad P\Big\{ \max_{1 \leq j \leq k} \xi_{l_j} \leq u \Big\} \leq \{\Phi(u)\}^k + c \sum_{1 \leq i < j \leq k} |r_{ij}| \exp\Big(-\frac{u^2}{1 + |r_{ij}|}\Big),$$

where $r_{ij} = \Lambda_{l_i l_j}$, and $c = c(\delta)$ is a constant independent of n, u and k, and we denote $\Phi(u) = \int_{-\infty}^{u} \frac{1}{\sqrt{2\pi}} \exp(-y^2/2)\, dy$.

Estimating an upper bound for the second term of the right hand side of (2.9), we can obtain the following lemma.

Lemma 2.6. *([3]) Let $\{\xi_j, j = 1, 2, \cdots, n\}, \delta, k$ and r_{ij} be as in Lemma 2.5. Further assume that the inequality $|r_{ij}| < |i - j|^{-\nu}$ holds for $\nu > 0$. Set $u = \{(2 - \eta)\log k\}^{1/2}$, where $0 < \eta < (1 - \delta)\nu/(1 + \nu + \delta)$. Then we have*

$$\sum := \sum_{1 \leq i < j \leq k} |r_{ij}| \exp\Big(-\frac{u^2}{1 + |r_{ij}|}\Big) \leq c\, k^{-\delta_0},$$

where $\delta_0 = \{\nu(1 - \delta) - \eta(1 + \delta + \nu)\}/\{(1 + \nu)(1 + \delta)\} > 0$ and c is a positive constant independent of n, u and k.

Now we are ready to prove Theorem 1.3:

Proof of Theorem 1.3. For any fixed $\theta > 1$, put

$$A_{k,j} = \{T : \theta^{k-1} \leq T \leq \theta^k,\ \theta^{j-1} \leq a_T \leq \theta^j\}, \quad k \geq 1,\ -\infty < j < \infty.$$

By the condition (iv), there exists $M > 0$ sufficiently large such that $T/a_T \geq M$ for all large $T > 0$, and let $k(\theta) = \big[\theta^{k-1-j}/M\big]$. Then

$$(2.10) \qquad \sup_{T \in A_{k,j}} \beta_1(T) \leq \theta\{2 \log k(\theta)\}^{1/2}$$

for k large enough. Let $\xi_i,\ i = 1, \cdots, d$, be normal random variables with means zero and let $a_i,\ i = 1, \cdots, d$, be positive numbers. Put $q = p/(p - 1)$ for $p > 1$, and define $\|\mathbf{b}\|_q = \big(\sum_{i=1}^{d} |b_i|^q\big)^{1/q}$ for $\mathbf{b} = (b_1, \cdots, b_d)$. It is well-known that

$$\Big(\sum_{i=1}^{d} a_i |\xi_i|^p\Big)^{1/p} = \sup_{\|\mathbf{b}\|_q \leq 1} \sum_{i=1}^{d} a_i^{1/p} b_i \xi_i$$

(cf. Lemma 2.3 in [13]). Setting $b_i = 1/\sqrt{d}$, $i = 1, \cdots, d$, we get

$$\frac{\|X^d(t + \theta^j) - X^d(t)\|}{\sigma(\theta^j)} \geq \frac{\sum_{i=1}^{d} \left(X_i(t + \theta^j) - X_i(t)\right)}{\sqrt{d}\,\sigma(\theta^j)}.$$

Let

$$Z_\theta(l) = \frac{\sum_{i=1}^{d} \left(X_i(lM\theta^j + \theta^j) - X_i(lM\theta^j)\right)}{\sqrt{d}\,\sigma(\theta^j)}, \quad 1 \leq l \leq k(\theta),$$

then $Z_\theta(l)$ are standard normal random variables. By the conditions (i) and (iv), there exists $j_0 > -\infty$ such that $j_0 \leq j \leq k/2$. Hence, by (2.10)

$$
\begin{aligned}
\liminf_{T \to \infty} \;&\sup_{0 \leq t \leq T} \sup_{0 \leq s \leq a_T} \frac{\|X^d(t + s) - X^d(t)\|}{\sigma(a_T)\,\beta_1(T)} \\[4pt]
&\geq \liminf_{k \to \infty} \min_{j_0 \leq j \leq k/2} \inf_{T \in A_{k,j}} \sup_{0 \leq t \leq T} \sup_{0 \leq s \leq a_T} \frac{\|X^d(t + s) - X^d(t)\|}{\sigma(a_T)\,\beta_1(T)} \\[4pt]
&\geq \liminf_{k \to \infty} \min_{j_0 \leq j \leq k/2} \sup_{0 \leq t \leq \theta^{k-1}} \frac{\|X^d(t + \theta^j) - X^d(t)\|}{\sigma(\theta^j)\,\theta(2 \log k(\theta))^{1/2}} \\[4pt]
&\geq \theta^{-1} \liminf_{k \to \infty} \min_{j_0 \leq j \leq k/2} \max_{1 \leq l \leq k(\theta)} \frac{\|X^d(lM\theta^j + \theta^j) - X^d(lM\theta^j)\|}{\sigma(\theta^j)\left(2 \log k(\theta)\right)^{1/2}} \\[4pt]
&\geq \theta^{-1} \liminf_{k \to \infty} \min_{j_0 \leq j \leq k/2} \max_{1 \leq l \leq k(\theta)} \frac{Z_\theta(l)}{\left(2 \log k(\theta)\right)^{1/2}} \\[4pt]
&=: \theta^{-1} J.
\end{aligned}
$$

(2.11)

The proof is completed if we show that

$$J \geq 1 \quad a.s. \tag{2.12}$$

since $\theta > 1$ is arbitrary. Using the elementary relation $ab = (a^2 + b^2 - (a - b)^2)/2$, it follows that, for $l \neq l'$,

$$
\begin{aligned}
\left| Cov\big(Z_\theta(l), Z(l')\big) \right| \\[4pt]
= \frac{1}{d\,\sigma^2(\theta^j)} \Bigg| \sum_{i=1}^{d} &E\{ X_i(Ml\theta^j + \theta^j) X_i(Ml'\theta^j + \theta^j) \\[4pt]
&- X_i(Ml\theta^j + \theta^j) X_i(Ml') - X_i(Ml\theta^j) X_i(Ml'\theta^j + \theta^j) \\[4pt]
&+ X_i(Ml\theta^j) X_i(Ml'\theta^j) \} \Bigg|
\end{aligned}
$$

(2.13)

$$\leq \frac{1}{2\,d\,\sigma^2(\theta^j)} \sum_{i=1}^{d} \left| \left(\sigma^2\big(|M(l-l')\theta^j + \theta^j|\big) - \sigma^2\big(|M(l-l')\theta^j|\big)\right) \right.$$

$$\left. - \left(\sigma^2\big(|M(l-l')\theta^j|\big) - \sigma^2\big(|M(l-l')\theta^j - \theta^j|\big)\right) \right|$$

$$= \frac{1}{2\,\sigma^2(\theta^j)} \left| \int_{|M(l-l')\theta^j|}^{|M(l-l')\theta^j+\theta^j|} d(\sigma^2(x)) - \int_{|M(l-l')\theta^j-\theta^j|}^{|M(l-l')\theta^j|} d(\sigma^2(x)) \right|.$$

Without loss of generality, assume that

$$|M(l-l')\theta^j - \theta^j| \leq |M(l-l')\theta^j| \leq |M(l-l')\theta^j + \theta^j|.$$

Hence, by Lemma 2.4 with $m = M$, $a = \theta^j$ and $b = l - l'$, the right hand side of (2.13) does not exceed

$$\frac{c\,\sigma^2(|M(l-l')\theta^j + \theta^j|)}{\sigma^2(\theta^j)\,|M(l-l') - 1|^2}.$$

Therefore, we obtain, by the regular variation of $\sigma(\cdot)$,

$$|r_{ll'}| := |Corr(Z_\theta(l), Z_\theta(l'))|$$

$$\leq c\,\frac{\sigma^2(|M(l-l')\theta^j + \theta^j|)}{\sigma^2(\theta^j)|M(l-l') - 1|^2}$$

$$\leq c\,\frac{1}{|M(l-l') - 1|^2}\left(\frac{\sigma^2(|M(l-l')\theta^j + \theta^j|)}{\sigma^2(\theta^j)}\right)$$

$$\leq c\,\frac{|M(l-l') + 1|^2}{|M(l-l') - 1|^2}|M(l-l') + 1|^{2\alpha-2}$$

$$< |l - l'|^{-\nu}$$

for M large enough, where $0 < \nu = 2 - 2\alpha$. Let us now apply Lemmas 2.5 and 2.6 for

$$\xi_l = Z_\theta(l), \quad |r_{ij}| = |r_{ll'}| < |l - l'|^{-\nu}, \quad 1 \leq l \leq k(\theta),$$

$$u = \{(2 - \eta)\log k(\theta)\}^{1/2}, \quad \eta = 2\varepsilon < \frac{(1-\delta)\nu}{1+\nu+\delta}.$$

Then we have

$$P\left\{\max_{1 \leq l \leq k(\theta)} \frac{Z_\theta(l)}{\{2\log k(\theta)\}^{1/2}} \leq \sqrt{1-\varepsilon}\right\}$$

$$\leq \{\Phi(u)\}^{k(\theta)} + c\,(k(\theta))^{-\delta_0} \leq \exp\{-c\,(k(\theta))^\varepsilon\} + c\,(k(\theta))^{-\delta_0}$$

$$\leq c(k(\theta))^{-\delta_0}.$$

Thus we obtain

$$P\left\{ \min_{j_0 \le j \le k/2} \max_{1 \le l \le k(\theta)} \frac{Z_\theta(l)}{\{2 \log k(\theta)\}^{1/2}} \le \sqrt{1 - \varepsilon} \right\}$$

$$\le \sum_{j_0 \le j \le k/2} P\left\{ \max_{1 \le l \le k(\theta)} \frac{Z_\theta(l)}{\{2 \log k(\theta)\}^{1/2}} \le \sqrt{1 - \varepsilon} \right\}$$

$$\le c \sum_{j_0 \le j \le k/2} \left(\theta^{k-1-j}/M \right)^{-\delta_0} \le c\,k\,\theta^{-k\delta_0/2}$$

and the Borel-Cantelli lemma gives (2.12).

<h2 align="center">REFERENCES</h2>

1. S. A. Book and T. R. Shore, *On large interval in the Csörgő-Révész theorem on increments of a Wiener process*, Z. Wahrsch. verw. Gebiete **46** (1978), 1-11.
2. Y. K. Choi and N. Kôno, *How big are the increments of a two-parameter Gaussian process?*, J. Theor. Probab. **12(1)** (1999), 105-129.
3. Y. K. Choi and S. H. Lee, *Some limit theorems on the increments of l^p-valued Gaussian processes*, to appear in Stochastic Analysis and Applications, Nova Science Publishers, Inc., 2001.
4. E. Csáki and P. Révész, *How big must be the increments of Wiener process?*, Acta Math. Acad. Sci. Hungar. **33** (1979), 37-49.
5. M. Csörgő and P. Révész, *Strong Approximations in probability and statistics*, New York, Academic Press, 1981.
6. C. M. Deo, *A note on increments of a Wiener process*, Technical Report, University of Ottawa, 1977.
7. X. Fernique, *Continuité des processus and Gaussiens*, C. R. Acad. Sci. Paris, t. **258** (1964), 6058-6060.
8. S. Hong, *Some results on increments of Gaussian processes*, Acta Math. Sinica **11A** (1990), 137-146.
9. N. Kôno, *The exact modulus of continuity for Gaussian processes taking values of a finite dimensional normed space*, Trends in Probability and Related Analysis, SAP'96, World Scientific, Singapore, 1996, 219-232.
10. M. R. Leadbetter, G. Lindgren and H. Rootzen, *Extremes and Related Properties of Random Sequences and Processes*, Springer-Verlag, New York, 1983.
11. Z. Y. Lin and C. R. Lu, *Strong Limit Theorems*, Kluwer Academic Publishers and Science Press, Hong Kong, 1992.
12. Q. M. Shao, *A note on increments of Wiener process*, J. Math. Sinica **6** (1986), 175-182 (in Chinese).
13. Q. M. Shao, *p-variation of Gaussian processes with stationary increments*, Studia Sci. Math. Hungar. **31** (1996), 237-247.
14. L. X. Zhang, *Some liminf results on increments of fractional Brownian motion*, Acta Math. Hungar. **71(3)** (1996), 215-240.

ON THE DEPENDENCE BETWEEN
COMPONENTS OF RANDOM VECTORS
WITH WEAKLY DEPENDENT CONDITIONS

TAE IL JEON[1]

ABSTRACT. In this article we summarize some results on a central limit theorem for a strictly stationary weakly dependent random vectors with some interlaced mixing conditions. We investigate the dependence between components of the random vector which is given as an asymptotic limit of an array of random vectors with interlaced mixing conditions. We discuss the cross covariance of the limiting vector process.

1. Introduction and Preliminary

Let $(\Omega, \mathcal{F}, P)$ be a probability space and $\mathcal{A}, \mathcal{B} \subset \mathcal{F}$ are sub σ-algebras of $\mathcal{F}$

Definition 1. Define the *strong mixing coefficient* by

$$(1) \qquad \alpha(\mathcal{A}, \mathcal{B}) = \sup_{A \in \mathcal{A}, B \in \mathcal{B}} |P(AB) - P(A)P(B)|$$

and the *maximal coefficient of correlation* by

$$(2) \qquad \rho(\mathcal{A}, \mathcal{B}) = \sup_{f \in L^2(\mathcal{A}), g \in L^2(\mathcal{B})} |\text{corr}(f, g)|.$$

Then the following inequality is elementary

$$\alpha(\mathcal{A}, \mathcal{B}) \leq \frac{1}{4}\rho(\mathcal{A}, \mathcal{B}).$$

Received December 29, 2000.

1991 *Mathematics Subject Classification.* 60F05.

Key Words and Phrases. Central limit theorem, weakly dependence.

[1] Department of Mathematics, Taejon University, Taejon, 300-716, Korea. Email: tijeon@dragon.taejon.ac.kr

For disjoint nonempty sets S and T, we use the following abbreviations:

$$\alpha(S,T) = \alpha(\sigma(X_n, n \in S), \sigma(X_n, n \in T)),$$
$$\rho(S,T) = \rho(\sigma(X_n, n \in S), \sigma(X_n, n \in T)),$$

where $\sigma(X_n, n \in S)$ denotes the σ-field generated by $\{X_n, n \in S\}$. For any positive integer n, define the following dependence coefficients for the given sequence of random variables $X = (X_n, n \in \mathbb{N})$

$$(3) \qquad \alpha(m) = \sup \alpha(S,T),$$
$$(4) \qquad \rho(m) = \sup \rho(S,T),$$

where, in both (3) and (4), the supremum is taken over all pairs of disjoint closed sets S and $T \subset \mathbb{N}$ with $\text{dist}(S,T) \geq m$.

Definition 2. A strictly stationary sequence $\{X_n\}$ is called α-*mixing* if $\alpha(n) \to 0$, where

$$\alpha(n) = \sup_k \alpha(\sigma(X_i, i \leq k), \sigma(X_i, i \geq k+n))$$

Note that the mixing coefficient $\alpha(n)$ is a measure of dependence between two sub σ-algebras $\sigma(X_i, i \leq k)$ and $\sigma(X_i, i \geq k+n)$. Let $\mathcal{F}_n^m = \sigma(X_n, X_{n+1}, \cdots, X_{n+m})$ and

$$\rho(n) = \sup\{|\text{corr}(f,g)| : f \in L_2(\mathcal{F}_{-\infty}^k), g \in L_2(\mathcal{F}_{k+n}^\infty)\}.$$

For any real n, define also

$$\alpha^*(n) = \sup \alpha(S,T), \quad S, T \subset \mathbb{R}, \quad \text{dist}(S,T) \geq n,$$
$$\rho^*(n) = \sup \rho(S,T), \quad S, T \subset \mathbb{R}, \quad \text{dist}(S,T) \geq n.$$

Comparing the definitions of $\alpha(n), \rho(n), \alpha^*(n)$ and $\rho^*(n)$, the following are obvious that $\alpha(n) \leq \alpha^*(n)$ and $\rho(n) \leq \rho^*(n)$.

Definition 3. A strictly stationary sequence $\{X_i\}_{i \in \mathbb{Z}}$ is called ρ-*mixing* if $\rho(n) \to 0$ as $n \to \infty$.

Let $\mathcal{F}_T = \sigma(X_i, i \in T)$, where T is a finite family of integers, and $\text{dist}(S,T) = \inf\{|s-t| : s \in S, t \in T\}$, where S and T are two nonempty finite subsets of $\mathbb{Z}$.

Definition 4. Let $\{X_i\}$ be a strictly stationary sequence. Then define

$$\alpha^*(n) = \sup \alpha(\mathcal{F}_T, \mathcal{F}_S), \quad \rho^*(n) = \sup \rho(\mathcal{F}_T, \mathcal{F}_S),$$

where these supremums are taken over all pairs of nonempty finite subsets S, T of $\mathbb{Z}$ such that $\mathrm{dist}(S, T) \geq n$.

Then we have, for $n \geq 1$,
(a) $\alpha^*(n) \leq \rho^*(n) \leq 2\pi\alpha^*(n)$,
(b) $\alpha(n) \leq \alpha^*(n)$,
(c) $\rho(n) \leq \rho^*(n)$.

2. Random Field

We may extend the definitions to a centered strictly stationary random field $X = (X_t, t \in \mathbb{R}^d)$ on $(\Omega, \mathcal{F}, P)$. The distance between any two disjoint nonempty subsets $S, T \subset \mathbb{R}^d$ will be denoted by $\mathrm{dist}(S, T) = \inf\{\|s - t\| : s \in S, t \in T\}$.

Theorem 5. (Bradley [2]) *Suppose that $d \geq 2$ and $X = (X_t, t \in \mathbb{R}^d)$ is a strictly stationary random field which is mixing and $r > 0$ is a real number. Then the following statements hold.*
(a) $\alpha(r) \leq \rho(r) \leq 2\pi\alpha(r)$.
(b) $\alpha(r) = \frac{1}{4} \iff \rho(r) = 1$.
(c) $\alpha^*(r) \leq \rho^*(r) \leq 2\pi\alpha^*(r)$.
(d) $\alpha^*(r) = \frac{1}{4} \iff \rho^*(r) = 1$.

By Theorem 5, the conditions $\alpha(r) \to 0$ and $\rho(r) \to 0$ are equivalent to each other for strictly stationary random fields $X = (X_t, t \in \mathbb{R}^d), d \geq 2$. Also we may say the conditions $\alpha^*(r) \to 0$ and $\rho^*(r) \to 0$ are equivalent to each other. Note that, for strictly stationary mixing random processes $\{X_t : t \in \mathbb{R}\}$ or $\{X_k : k \in \mathbb{Z}\}$, (a) and (b) are not true in general. Bradley [2] proved, for the strictly stationary sequences, that the condition $\alpha^*(n) \to 0$ as $n \to \infty$ contains enough information to assure the CLT without any additional rate or moment conditions higher than 2. We state the result:

Theorem 6. *If $(X_k, k \in \mathbb{Z})$ is a strictly stationary sequence of real centered square integrable random variables such that $\sigma_n^2 = \mathrm{var}(\sum_{i=1}^n X_i) \to \infty$ and $\alpha^*(n) \to 0$ as $n \to \infty$, then*

$$\frac{\sum_{i=1}^n X_i}{\sigma_n} \xrightarrow{d} \mathcal{N}(0, 1) \quad \text{as } n \to \infty.$$

Consider some notations for d-dimensional block sum. Suppose that

$$j = (j_1, \cdots, j_d), \quad l = (l_1, \cdots, l_d)$$

are elements in $\mathbb{Z}^d$ such that $j_u \le l_u$, $u = 1, \cdots, d$. Denote

$$S(X : j_1, \cdots, j_d; l_1, \cdots, l_d) = \sum_k X_k,$$

where the sum is taken over all $k = (k_1, \cdots, k_d) \in \mathbb{Z}^d$ such that $j_u \le k_u \le l_u$, $u = 1, \cdots, d$. For positive integers $l_1, \cdots, l_d$, we use the simple notations

$$\begin{aligned}
S(X : l) &= S(X : l_1, \cdots, l_d) \\
&= S(X : 1, \cdots, 1; l_1, \cdots, l_d).
\end{aligned}$$

Considering the asymptotic normality of $S(X : n)$ as $n \in \mathbb{N}^d$ becomes large in the sense of the usual Euclidean norm, Bradley have proved a CLT for the strictly stationary random fields with finite second moments and the corresponding unrestricted ρ-mixing conditions. Also no mixing rate is assumed.

Theorem 7. *Suppose that d is a positive integer and $X = (X_k, k \in \mathbb{Z}^d)$ is a centered real strictly stationary random field such that $0 < EX_0^2 < \infty, \rho^*(r) \to 0$ as $r \to \infty$, and the continuous spectral density f of X on T^d satisfies $f(1, \cdots, 1) > 0$. Then, as $\|n\| \to \infty$, it follows that $\|S(X : n)\|_2 \to \infty$ and*

$$\frac{S(X : n)}{\|S(X : n)\|_2} \xrightarrow{d} \mathcal{N}(0, 1).$$

Let d be a positive integer. Suppose that $X = (X_k, k \in \mathbb{Z}^d)$ is a centered strictly stationary random field on a probability space $(\Omega, \mathcal{F}, P)$. Let us denote the usual Euclidean norm of a vector $k = (k_1, \ldots, k_d) \in \mathbb{Z}^d$ by $\|k\|$ and $\||k\|| = k_1 \cdots k_d$. The distance between two disjoint nonempty subsets $S, T \subset \mathbb{Z}^d$ will be denoted by

$$\text{dist}(S, T) = \min_{j \in S, k \in T} \|j - k\|.$$

Then we have the following result:

Theorem 8. *Let $X = \{X_n, n \in \mathbb{Z}^d\}$ be a strictly stationary random field which is centered and has finite second moments. Assume that*
    (a) $\alpha(n) \to 0$ *as* $\|n\| \to \infty$,
    (b) $\lim_{\|n\| \to \infty} \rho^*(n) < 1$,
    (c) $\sigma_n^2 = \text{var} S(X : n) \to \infty$ *as* $\|n\| \to \infty$.

Then

$$\liminf \frac{\sigma_n^2}{|||n|||} > 0$$

and

$$\frac{S(X:n)}{\sigma_n} \xrightarrow{d} \mathcal{N}(0,1) \quad \text{as } \|n\| \to \infty.$$

3. Triangular Array

Consider a triangular arrays of strongly mixing random variables $\{\xi_{ni}, 1 \leq i \leq k_n\}$, where $k_n \to \infty$. We shall define the following

$$(5) \qquad \overline{\alpha}_{nk} = \sup_{s \geq 1} \alpha(\sigma(\xi_{ni}, i \leq s), \sigma(\xi_{nj}, j \geq s+k))$$

and $\overline{\alpha}_k = \sup_n \overline{\alpha}_{nk}$. The array will be called *strongly mixing* if $\overline{\alpha}_k \to 0$ as $k \to \infty$. Similarly, we define

$$(6) \qquad \overline{\rho}_{nk}^* = \sup_k \rho(\sigma(\xi_{ni}, i \in T), \sigma(\xi_{nj}, j \in S)),$$

where $T, S \subset \{1, 2, \cdots, k_n\}$ are nonempty subsets with $\mathrm{dist}(T, S) \geq k$ and

$$(7) \qquad \overline{\rho}_k^* = \sup_n \overline{\rho}_{nk}^*.$$

Theorem 9. (Peligrad [4]) *Let $\{\xi_{ni}; 1 \leq i \leq k_n\}$ be a triangular array of centered random variables, which is strongly mixing and have finite second moments. Assume that $\lim_{k \to \infty} \overline{\rho}_k^* < 1$. Denote by $\sigma_n^2 = \mathrm{var}(\sum_{i=1}^{k_n} \xi_{ni})$ and assume that*

$$(8) \qquad \sup_n \frac{1}{\sigma_n^2} \sum_{i=1}^{k_n} E\xi_{ni}^2 < \infty$$

and, for any $\varepsilon > 0$,

$$(9) \qquad \frac{1}{\sigma_n^2} \sum_{i=1}^{k_n} E\xi_{ni}^2 I(|\xi_{ni}| > \varepsilon \sigma_n) \to 0 \quad \text{as } n \to \infty.$$

Then

$$(10) \qquad \frac{\sum_{i=1}^{k_n} \xi_{ni}}{\sigma_n} \xrightarrow{d} \mathcal{N}(0,1) \quad \text{as } n \to \infty.$$

Note that the stationarity condition is not assumed in Theorem 9. When we assume the stationary condition on a sequence of random variables, we can formulate a theorem as follows (Peligrad [4]):

Theorem 10. *Suppose that $\{X_k, k \in \mathbb{N}\}$ is a strongly mixing strictly stationary sequence of random variables which is centered and has finite second moments. Assume that $\lim_{n \to \infty} \rho^*(n) < 1$ and $\sigma_n^2 \to \infty$. Then*

$$(11) \qquad \liminf \frac{\sigma_n^2}{n} > 0,$$

$$(12) \qquad \frac{\sum_{k=1}^{n} X_k}{\sigma_n} \xrightarrow{d} \mathcal{N}(0,1) \quad \text{as } n \to \infty.$$

We consider a triangular array of two-dimensional random variates

$$\{\xi_{ni} \mid 1 \leq i \leq k_n\} = big\{ \left(\xi_{ni}^{(1)}, \xi_{ni}^{(2)} \right) \mid 1 \leq i \leq k_n \}$$

such that $\xi_{ni}^{(j)}, j = 1, 2$, satisfy some interlaced mixing conditions. We will investigate the independence between the componentwise limiting distributions of the given array.

4. Triangular Array of Random Vectors

Consider a triangular array of bivariate random variables $\xi_{ni} = \left(\xi_{ni}^{(1)}, \xi_{ni}^{(2)} \right)$, where $1 \leq i \leq k_n$ and $k_n \to \infty$. Here we refer to the definitions used in Peligrad [4]. Define, for $j = 1, 2$,

$$\bar{\alpha}_{nk}^{(j)} = \sup_{s \geq 1} \alpha \left(\sigma \left(\xi_{ni}^{(j)}, i \leq s \right), \sigma \left(\xi_{ni}^{(j)}, i \geq s + k \right) \right)$$

and $\bar{\alpha}_k^{(j)} = \sup_n \bar{\alpha}_{nk}^{(j)}$. The array $\{\xi_{ni} | 1 \leq i \leq k_n\}$ will be called *strongly mixing* if, for each $j = 1, 2$, $\lim_{k \to \infty} \bar{\alpha}_k^{(j)} = 0$. In order to properly define the corresponding ρ-mixing coefficients for the array, we have the following definition:

$$\bar{\rho}_{nk}^{(j)*} = \sup_{k \geq 1} \rho \left(\sigma \left(\xi_{ni}^{(j)}, i \in T \right), \sigma \left(\xi_{ni}^{(j)}, i \in S \right) \right),$$

where $T, S \subset \{1, 2, \cdots, k_n\}$ are nonempty subsets with $\text{dist}(T, S) \geq k$ and $\bar{\rho}_k^{(j)*} = \sup_n \bar{\rho}_{nk}^{(j)*}$.

Theorem 11. *Let $\left\{ \xi_{ni} = \left(\xi_{ni}^{(1)}, \xi_{ni}^{(2)} \right) | 1 \leq i \leq k_n \right\}$ be a triangular array of centered bivariate random variables which is strongly mixing componentwise and*

$E(\xi_{ni}^{(j)})^2 < \infty$ for each $j = 1, 2$. Assume that $\lim_{k \to \infty} \bar{\rho}_k^{(j)*} < 1$ for each $j = 1, 2$. Denote, for each $j = 1, 2$, by $\left(\sigma_n^{(j)}\right)^2 = \mathrm{var}\left(\sum_{i=1}^{k_n} \xi_{ni}^{(j)}\right)$ and assume that

$$\sup_n \frac{1}{\left(\sigma_n^{(j)}\right)^2} \sum_{i=1}^{k_n} E\left(\xi_{ni}^{(j)}\right)^2 < \infty$$

and, for every $\varepsilon > 0$,

$$\frac{1}{\left(\sigma_n^{(j)}\right)^2} \sum_{i=1}^{k_n} E\left(\xi_{ni}^{(j)}\right)^2 I\left(\left|\xi_{ni}^{(j)}\right| > \varepsilon \sigma_n^{(j)}\right) \to 0 \quad \text{as } n \to \infty.$$

Then, for each $j = 1, 2$,

$$\frac{\sum_{i=1}^{k_n} \xi_{ni}^{(j)}}{\sigma_n^{(j)}} \xrightarrow{d} \mathcal{N}(0, 1) \quad \text{as } n \to \infty,$$

where d stands for the convergence in distribution.

5. Cross Covariance of Vector Process and CLT

Throughout this section, we assume that the conditions in Theorem 11 in section 1 hold. Note that the results stated in the previous section guarantee the componentwise CLT for the given array of vector process with componentwise mixing conditions. What we are interested in is to investigate the limiting behavior of the array of vector process $\{\xi_{ni} | 1 \leq i \leq k_n\}$. For each $j = 1, 2$, let $\zeta_{ni}^{(j)} = \xi_{ni}^{(j)} / \sigma_n^{(j)}$. To investigate the limiting behavior, it is necessary to see the limiting behavior of the cross covariance of the sum of sequence of normalized vector process $\{\zeta_{ni} | 1 \leq i \leq k_n\}$. It is natural to compute the cross covariance of the vector process $\{\zeta_{ni} | 1 \leq i \leq k_n\}$. We study whether the sequence of vector process

$$(13) \qquad \left(\sum_{i=1}^{k_n} \zeta_{ni}^{(1)}, \sum_{i=1}^{k_n} \zeta_{ni}^{(2)} \right)$$

converges in distribution to a bivariate normal vector process (Z_1, Z_2), where $Z_1 \sim \mathcal{N}(0, 1), Z_2 \sim \mathcal{N}(0, 1)$. To do this, we need to sketch the proof of Theorem 11. (See Peligrad [4] for the detail of the proof). We can construct a sequence of positive numbers ε_n such that

$$\sum_{i=1}^{k_n} E\big(\zeta_{ni}^{(j)}\big)^2 I\big(|\zeta_{ni}^{(j)}| > \varepsilon_n\big) \to 0 \quad \text{as } n \to \infty.$$

Truncating at the level ε_n, define, for each $j = 1, 2$,

$$\eta_{ni}^{(j)} = \zeta_{ni}^{(j)} I\big(|\zeta_{ni}^{(j)}| \le \varepsilon_n\big) - E\zeta_{ni}^{(j)} I\big(|\zeta_{ni}^{(j)}| \le \varepsilon_n\big)$$

and

$$\gamma_{ni}^{(j)} = \zeta_{ni}^{(j)} I\big(|\zeta_{ni}^{(j)}| > \varepsilon_n\big) - E\zeta_{ni}^{(j)} I\big(|\zeta_{ni}^{(j)}| > \varepsilon_n\big).$$

Note that, for each $j = 1, 2$, $\zeta_{ni}^{(j)} = \eta_{ni}^{(j)} + \gamma_{ni}^{(j)}$. Then we can show that

$$(14) \qquad \operatorname{var}\left(\sum_{i=1}^{k_n} \gamma_{ni}^{(j)}\right) \to 0 \quad \text{as } n \to \infty.$$

By the argument above, we can rewrite the statement in Theorem 11 into the following:

Theorem 12. *Let $\{\eta_{ni} | 1 \le i \le k_n\}$ be an array of bivariate random variables with zero means and finite second moments, and satisfies, for each $j = 1, 2$,*

$$|\eta_{ni}^{(j)}| \le 2\varepsilon_n, \quad \text{where } \varepsilon_n > 0,$$

$$(\sigma_n^{(j)})^2 = \operatorname{var}\left(\sum_{i=1}^{k_n} \eta_{ni}^{(j)}\right) \to 1 \quad \text{as } n \to \infty$$

and

$$\sup_n \sum_{i=1}^{k_n} \operatorname{var}\big(\eta_{ni}^{(j)}\big) < \infty.$$

Then, for each $j = 1, 2$,

$$\sum_{i=1}^{k_n} \eta_{ni}^{(j)} \xrightarrow{d} \mathcal{N}(0, 1) \quad \text{as } n \to \infty.$$

Here we apply the blocking procedure to divide the sequence of random variables into big blocks and small blocks. Since we constructed a sequence $\{\varepsilon_n\}$, we construct a sequence of integers $\{q_n\}$ such that the following conditions satisfied

$$q_n \to \infty, \quad q_n \varepsilon_n \to 0, \quad q_n \bar{\alpha}^{(j)}\big([\varepsilon_n^{-1}]\big) \to 0 \quad \text{as } n \to \infty,$$

where $\bar{a}^{(j)}(n) = \bar{a}_n^{(j)}$. For each $n \in \mathbb{N}$, define the integers by the following steps:

$$m_0 = 0$$

and, for $p = 0, 1, 2, \cdots$, let

$$m_{2p+1} = \min\left\{ m \mid m > m_{2p}, \sum_{i=m_{2p}+1}^{m} \operatorname{var}(\eta_{ni}^{(j)}) \geq q_n^{-1} \right\},$$

$$m_{2p+2} = m_{2p+1} + [\varepsilon_n^{-1}].$$

We classify the groups specified by the argument above. Let

$$I_p = \{k \mid m_{2p} < k \leq m_{2p+1}\},$$
$$J_p = \{k \mid m_{2p+1} < k \leq m_{2p+2}\}$$

for $p = 0, 1, 2, \cdots$. Processing the steps above, we have ℓ_n blocks of indexes I_p and J_p, respectively, $p = 0, 1, 2, \cdots, \ell_n$. Denote

$$Y_{np} = \sum_{i \in I_p} \eta_{ni}, \quad Z_{np} = \sum_{i \in J_p} \eta_{ni}, \quad 0 \leq p \leq \ell_n.$$

By an appropriate argument, we can show that

$$(15) \qquad \operatorname{var}\left(\sum_{p=1}^{\ell_n} Z_{np} \right) \to 0 \quad \text{as } n \to \infty,$$

which means that $\sum_{p=1}^{\ell_n} Z_{np}$ is negligible for the convergence in distribution. Moreover, we get

$$(16) \qquad \lim_{n \to \infty} \operatorname{var}\left(\sum_{p=1}^{\ell_n} Y_{np} \right) = 1$$

and there exist two constants $0 < K_1 < K_2$ and $N \in \mathbb{N}$ such that

$$0 < K_1 < \sum_{p=1}^{\ell_n} \operatorname{var} Y_{np} < K_2, \quad n \geq N.$$

Let $a_n = \left(\sum_{p=1}^{\ell_n} \operatorname{var} Y_{np} \right)^{1/2}$. Then we may assume that $\{Y_{np} \mid 0 \leq p \leq \ell_n\}$ is an array of independent random variables.

Finally, we may show that $\{a_n^{-1} Y_{np} \mid 0 \le p \le \ell_n\}$ satisfies the CLT. Therefore, the proof of Theorem 11 is completed. As mentioned in the proof of Theorem 11 in the section 1, we have known how the big blocks are constructed and that they act like an array of independent random variables. We will see what the big blocks say about the dependence between the components in the limiting distribution. So it is worth investigating the limit of the following sequence of cross covariances

$$(17) \qquad \mathrm{Cov}\left(\sum_{i=1}^{k_n} \zeta_{ni}^{(1)}, \sum_{i=1}^{k_n} \zeta_{ni}^{(2)} \right)$$

for the given array of random vectors $\{\xi_{ni} = (\xi_{ni}^{(1)}, \xi_{ni}^{(2)}) \mid 1 \le i \le k_n\}$. By a simple argument, we have the following fact, which states the given mixing conditions on each component of the sequence of vector process give at least the uniform boundedness of the sequence of cross covariances.

Theorem 13. *Let $\{\eta_{ni} \mid 1 \le i \le k_n\}$ be an array of bivariate random variables with zero means and finite second moments and satisfy, for each $j = 1, 2$,*

$$|\eta_{ni}^{(j)}| \le 2\varepsilon_n, \quad \text{where } \varepsilon_n > 0,$$

$$(\sigma_n^{(j)})^2 = \mathrm{var}\left(\sum_{i=1}^{k_n} \eta_{ni}^{(j)} \right) \to 1 \quad \text{as } n \to \infty$$

and

$$\sup_n \sum_{i=1}^{k_n} \mathrm{var}\, \eta_{ni}^{(j)} < \infty.$$

Then there is a number M such that

$$\left| \mathrm{Cov}\left(\sum_{i=1}^{k_n} \eta_{ni}^{(1)}, \sum_{i=1}^{k_n} \eta_{ni}^{(2)} \right) \right| \le M, \quad n \ge 1.$$

If the sequence in (17) converges to a number ρ, then we have that the array of random vectors in (13) converges in distribution to a bivariate normal distribution (Z_1, Z_2) with covariance matrix

$$\Sigma = \begin{pmatrix} 1 & \rho \\ \rho & 1 \end{pmatrix}.$$

To investigate the convergence of (17), for $j = 1, 2$, let $A_n^{(j)} = \sum_{i=1}^{k_n} \xi_{ni}^{(j)}$. Then $A_n^{(j)} = \sigma_n^{(j)} \zeta_n^{(j)}, j = 1, 2$. Consider the following sequence

$$(18) \qquad E\left[\left(\sum_{i=1}^{k_n} \zeta_{ni}^{(1)}\right)\left(\sum_{i=1}^{k_n} \zeta_{ni}^{(2)}\right)\right].$$

Then we may show that (18) is a Cauchy sequence and hence converges. We state the result as the following:

Proposition 14. *Suppose that, for each $j = 1, 2$,*

$$(19) \qquad E\left[\left(\sum_{i=1}^{k_n} \zeta_{ni}^{(j)}\right)\left(\sum_{i=1}^{k_m} \zeta_{mi}^{(j)}\right)\right] \to 1 \quad as \ n, m \to \infty.$$

Then there is a number ρ such that

$$E\left[\left(\sum_{i=1}^{k_n} \zeta_{ni}^{(1)}\right)\left(\sum_{i=1}^{k_n} \zeta_{ni}^{(2)}\right)\right] \to \rho \quad as \ n \to \infty.$$

This indicates that the limiting behavior of the auto covariance of the sum of each array of $\{\zeta_{ni}^{(j)} | 1 \leq i \leq k_n\}, j = 1, 2$ determines the limiting behavior of the cross covariance of the normalized vector process $\{\zeta_{ni} | 1 \leq i \leq k_n\}$.

Theorem 15. *The necessary and sufficient condition that (19) hold is, for each $j = 1, 2$,*

$$(20) \qquad E\left(\sum_{p=1}^{\ell_n} Y_{np}^{(j)}\right)\left(\sum_{p=1}^{\ell_m} Y_{mp}^{(j)}\right) \to 1 \quad as \ n, m \to \infty.$$

Theorem 15 indicates that the limiting behavior of the auto covariance of the sequence of sum of big blocks determines the limiting behavior of the cross covariance of the normalized vector process $\{\zeta_{ni} | 1 \leq i \leq k_n\}$. Note that we have only the uniform boundedness of the sequence in (19) or (20). But it does not guarantee the convergence of the sequence in (20). It seems not to be easy to find some sufficient conditions to have the convergence of (19). It might help finding sufficient conditions if we add some mixing conditions between components of an array of vector process. We leave it as an open question. To force the convergence of it, we need a strong condition, for instance, convergence in probability. Therefore, we assume that, for each $j = 1, 2$, $\{\sum_{i=1}^{k_n} \zeta_{ni}^{(j)}\}$ is Cauchy in probability. Then (19) holds true since the sequence of second moments of each component converges to the second moment of the normal distribution, that is, 1. By those argument, we have the following:

Theorem 16. *Suppose that, for each $j = 1, 2$, $\left\{ \sum_{i=1}^{k_n} \zeta_{ni}^{(j)} \right\}$ is Cauchy in probability. Then there exists a number $0 \le \rho \le 1$ such that*

$$\lim_{n \to \infty} \mathrm{Cov}\left(\sum_{i=1}^{k_n} \zeta_{ni}^{(1)}, \sum_{i=1}^{k_n} \zeta_{ni}^{(2)} \right) = \rho$$

and hence

$$\left(\sum_{i=1}^{k_n} \zeta_{ni}^{(1)}, \sum_{i=1}^{k_n} \zeta_{ni}^{(2)} \right) \xrightarrow{d} (Z_1, Z_2),$$

where (Z_1, Z_2) is a bivariate normal distribution $\mathcal{N}_2(\mu, \Sigma)$. Here the expectation and the covariance matrix are as like

$$\mu = (EZ_1, EZ_2) = (0, 0), \quad \Sigma = \begin{pmatrix} 1 & \rho \\ \rho & 1 \end{pmatrix}.$$

References

1. R. C. Bradley, *On the spectral density and asymptotic normality of weakly dependent random fields*, J. Theor. Prob. **5** (1992), 355–373.
2. R. C. Bradley Equivalent mixing conditions for random fields, Ann. Prob. **21(4)** (1993), 1921–1926.
3. W. Bryc and W. Smolenski (1993) Moment conditions for almost sure convergence of weakly correlated random variables, Proc. Amer. Math. Soc. **119(2)** (1993), 629–635.
4. M. Peligrad, *On the asymptotic normality of sequences of weakly dependent random variables*, J. Theo. Prob. **9(3)** (1996), 703–716.

MODULI OF CONTINUITY ON THE INCREMENTS OF l^∞-VALUED GAUSSIAN PROCESSES

SHIN YOUNG KANG[1] AND YONG KAB CHOI[1]

ABSTRACT. Moduli of continuity on the increments of an l^∞-valued Gaussian process are obtained via estimating upper bounds of large and small deviation probabilities on the suprema of the Gaussian process.

1. Introduction and Results

Let $\{Y(t), -\infty < t < \infty\} = \{X_k(t), -\infty < t < \infty\}_{k=1}^\infty$ be a sequence of continuous Gaussian processes with stationary increments. Csörgő, Lin and Shao [3] and Lin and Qin [7] established moduli of continuity and large increment results for l^∞-valued Gaussian processes of $Y(\cdot)$, respectively.

In this paper we modify and generalize the moduli of continuity for $Y(\cdot)$ in [3] under weaker conditions. We always assume the following statements: Let $\{X_k(t), 0 \le t < \infty\}_{k=1}^\infty$ be a sequence of real-valued separable, centered Gaussian processes with $X_k(0) = 0$ and stationary increments

$$(1.1) \qquad E\{X_k(t) - X_k(s)\}^2 = \sigma_k^2(|t - s|), \quad 0 < |t - s| < 1.$$

Assume that $\sigma_k(h)$ are positive non-decreasing continuous and regularly varying functions of $h > 0$ with exponents α_k ($0 < \alpha_k < 1$) at 0 and ∞, that is, $\sigma_k(h) = h^{\alpha_k} L_k(h)$, where $L_k(h)$ are slowly varying functions at 0 and ∞. Put $\sigma^*(h) = \sup_{k \ge 1} \sigma_k(h)$. Let $Y(t) = (X_1(t), X_2(t), \cdots)$, $0 \le t < \infty$, be an infinite dimensional Gaussian process with l^∞-norm $\|\cdot\|_\infty$ defined by $\|Y(t)\|_\infty = \sup_{k \ge 1} |X_k(t)|$. Further, assume that

$$(i) \qquad \sigma^{*2}(h)/\sigma_k^2(h) \ge 1 + \log k, \quad k \ge 1.$$

Received November 9, 2000.

1991 *Mathematics Subject Classification.* 60F15, 60G15.

Key Words and Phrases. l^∞-norm, regurarly varying function, Gaussian process.

This work was supported by Korea Research Foundation Grant (KRF-99-005-D00003).

[1] Department of Mathematics, Gyeongsang National University, Chinju 660-701, Korea. Email: ksyoung9@hanmail.net and mathykc@nongae.gsnu.ac.kr

Remark 1. The condition (i) implies that there exists a constant $A > 0$ such that $\sum_{k=1}^{\infty} \sigma_k^A(h_0) \leq \infty$ for some $0 < h_0 < 1$. Thus (i) is weaker than the condition (2.2): $\sum_{k=1}^{\infty} \sigma_k^A(h_0) < \infty$ in [3].

For example, taking $\sigma_k(h) = h^\alpha/k$ $(0 < \alpha < 1)$, $k \geq 1$, the condition (i) is satisfied and we get $\sum_{k=1}^{\infty} \sigma_k^A(h_0) < \infty$ for some $A > 0$, which is the condition (2.2) in [3]. But if we take $\sigma_k(h) = h^\alpha/(1 + 4\log k)$ $(0 < \alpha < 1)$, $k \geq 1$, then (i) is satisfied and $\sum_{k=1}^{\infty} \sigma_k^A(h_0) = \infty$ for any $A > 0$, and the latter is not the case of (2.2).

The main results are as follows:

Theorem 1.1. *We have*

$$(1.2) \qquad \limsup_{h \to 0} \; \sup_{0 \leq t \leq 1} \; \sup_{0 \leq s \leq h} \frac{\|Y(t+s) - Y(t)\|_\infty}{\sigma^*(h)(2\log(1/h))^{1/2}} \leq 1 \quad \text{a.s.}$$

Theorem 1.2. *Further assume that $X_k(\cdot)$, $k = 1, 2, \cdots$ are independent. Let y_h be the solution of equation*

$$(\text{ii}) \qquad \sum_{k=1}^{\infty} (h y_h)^{\sigma^{*2}(h)/\sigma_k^2(h)} = h$$

and let $\sigma_k^2(h)$ be twice differentiable which satisfies

$$(\text{iii}) \qquad \left| \frac{d^2 \sigma_k^2(h)}{dh^2} \right| \leq c \frac{\sigma_k^2(h)}{h^2}$$

for some $c > 0$, then we have

$$(1.3) \qquad \liminf_{h \to 0} \; \sup_{0 \leq t \leq 1} \; \sup_{0 \leq s \leq h} \frac{\|Y(t+s) - Y(t)\|_\infty}{\sigma^*(h)(2\log(1/h y_h))^{1/2}} \geq 1 \quad \text{a.s.}$$

and

$$(1.4) \qquad \limsup_{h \to 0} \; \sup_{0 \leq t \leq 1} \frac{\|Y(t+h) - Y(t)\|_\infty}{\sigma^*(h)(2\log(1/h y_h))^{1/2}} \geq 1 \quad \text{a.s.}$$

Remark 2. For $k \geq 1$, let $\sigma_k(h) = h^\alpha/\sqrt{1 + \log k}$ $(0 < \alpha < 1)$ and $y_h = \frac{1}{h}(2^{-k}h)^{1/(1+\log k)}$, then $\sum_{k=1}^{\infty} \sigma_k^A(h) = \infty$ for any $A > 0$ and (ii) is satisfied.

Combining Theorems 1.1 and 1.2, we get the following:

Corollary 1.3. *Under the assumptions of Theorem 1.2, we have*

$$\lim_{h\to 0}\sup_{0\le t\le 1}\sup_{0\le s\le h}\frac{\|Y(t+s)-Y(t)\|_\infty}{\sigma^*(h)(2\log(1/hy_h))^{1/2}}=1\quad a.s.,$$

$$\lim_{h\to 0}\sup_{0\le t\le 1}\sup_{0\le s\le h}\frac{\|Y(t+s)-Y(t)\|_\infty}{\sigma^*(h)(2\log(1/h))^{1/2}}=1\quad a.s.,$$

$$\limsup_{h\to 0}\sup_{0\le t\le 1}\frac{\|Y(t+h)-Y(t)\|_\infty}{\sigma^*(h)(2\log(1/h))^{1/2}}=1\quad a.s.,$$

$$\limsup_{h\to 0}\sup_{0\le t\le 1}\frac{\|Y(t+h)-Y(t)\|_\infty}{\sigma^*(h)(2\log(1/hy_h))^{1/2}}=1\quad a.s.$$

2. The Proofs

We shall accomplish the proofs of Theorems 1.1 and 1.2 via the following several lemmas. Let T be a compact subset of $\mathbb{R}^N$ with the Euclidean norm $\|\cdot\|$, and let $\{X_i(t),t\in T\}_{i=1}^\infty$ be a sequence of real-valued separable, centered Gaussian processes. Assume that $Y(\mathbf{t})=(X_1(\mathbf{t}),X_2(\mathbf{t}),\cdots)$, $\mathbf{t}\in T$, is an infinite dimensional Gaussian process with $\|\cdot\|_\infty$. Suppose that

$$
\begin{aligned}
&0<\Gamma_i:=\sup_{t\in T}\{E(X_i(\mathbf{t}))^2\}^{1/2}<\infty,\quad \Gamma:=\sup_{i\ge 1}\Gamma_i,\\
&E\{X_i(\mathbf{t})-X_i(\mathbf{s})\}^2\le\varphi_i^2(\|\mathbf{t}-\mathbf{s}\|)\quad\text{and}\quad \varphi^*(h)=\sup_{i\ge 1}\varphi_i(h),
\end{aligned}
$$

(2.1)

where $\varphi_i(\cdot)$ are nondecreasing continuous functions. The following lemma is another version of Fernique lemma [4] on an l^∞-valued Gaussian process, whose proof is similar to that of Lemma 2.3 in [2].

Lemma 2.1. *For $\lambda>0$ and $\mathcal{B}>(4\sqrt{2}+4)\sqrt{N}$, there exists a constant $c>0$ such that*

$$P\left\{\sup_{t\in T}\|Y(\mathbf{t})\|_\infty\ge x\left(\Gamma+\mathcal{B}\int_0^\infty\varphi^*(\sqrt{N}\lambda 2^{-y^2})\,dy\right)\right\}\le c\,\frac{m(T)}{\lambda^N}\exp(-x^2/2)$$

for all $x>0$, where $m(T)$ is the Lebesgue measure of T.

Let us estimate an upper bound of the following large deviation probabilities by using Lemma 2.1:

Lemma 2.2. *For any $\varepsilon>0$, there exists a positive constant c_ε depending only on ε such that*

$$P\left\{\sup_{0\le t\le 1}\sup_{0\le s\le h}\frac{\|Y(t+s)-Y(t)\|_\infty}{\sigma^*(h)}\ge x\right\}\le c_\varepsilon\frac{1}{h}\exp\left(-\frac{x^2}{2+\varepsilon}\right)$$

for all $x > 0$.

Proof. Let $\mathbb{D} = \{(t,s) : 0 \le t \le 1, \ 0 \le s \le h\}$ be a two-parameter set. In order to apply Lemma 2.1, we set, for $k \ge 1$,

$$Z_k(t,s) = \frac{X_k(t+s) - X_k(t)}{\sigma^*(h)}, \quad (k,s) \in \mathbb{D},$$

and

$$\varphi_k(z) = \frac{2\sigma_k(\sqrt{2}z)}{\sigma^*(h)}, \quad z > 0.$$

Clearly, for $k \ge 1$, we have $E\{Z_k(t,s)\} = 0$ and $\Gamma = 1$. Letting $u = (t',s')$ and $v = (t'',s'')$, it follows from the relation $(a-b)^2 \le 2(a^2+b^2)$ that

$$
\begin{aligned}
&E\{Z_k(u) - Z_k(v)\}^2 \\
&\le \frac{2}{\sigma^{*2}(h)} E\{(X_k(t'+s') - X_k(t''+s''))^2 + (X_k(t') - X_k(t''))^2\} \\
&= \frac{2}{\sigma^{*2}(h)} \{\sigma_k^2(|(t'+s') - (t''+s'')|) + \sigma_k^2(|t'-t''|)\} \\
&\le \frac{4}{\sigma^{*2}(h)} \sigma_k^2(\sqrt{2}\sqrt{|t'-t''|^2 + |s'-s''|^2}) \\
&= \varphi_k^2(\|u-v\|).
\end{aligned}
$$

On the other hand, for any $\varepsilon > 0$, we can find a small constant $c_\varepsilon > 0$ such that

$$\mathcal{B} \int_0^\infty \varphi^*(\sqrt{2}\,c_\varepsilon\, h\, 2^{-y^2})\, dy < \varepsilon/8$$

for all $0 < h < 1$, where $\mathcal{B} > (4\sqrt{2}+4)\sqrt{N}$. Let $Z(t,s) = (Z_1(t,s), Z_2(t,s), \cdots)$. Let $x = u(1 + \varepsilon/8)$, $u > \sqrt{2}$. It follows from Lemma 2.1 that

$$
\begin{aligned}
&P\left\{ \sup_{0 \le t \le 1} \sup_{0 \le s \le h} \frac{\|Y(t+s) - Y(t)\|_\infty}{\sigma^*(h)} \ge x \right\} \\
&\le P\left\{ \sup_{(t,s) \in \mathbb{D}} \|Z(t,s)\|_\infty \ge u(1 + \mathcal{B} \int_0^\infty \varphi^*(\sqrt{2}c_\varepsilon h 2^{-y^2})dy) \right\} \\
&\le c \frac{h}{(c_\varepsilon h)^2} e^{-u^2/2} \le c_\varepsilon \frac{1}{h} \exp\left(-\frac{x^2}{2+\varepsilon}\right).
\end{aligned}
$$

In case $0 < x \le \sqrt{2}$, the result follows immediately if we take c_ε large enough.

Proof of Theorem 1.1. By Lemma 2.2, for given $\varepsilon > 0$, there exists a constant $c = c(\varepsilon) > 0$ such that

(2.3)
$$P\left\{ \sup_{0 \le t \le 1} \sup_{0 \le s \le h} \frac{\|Y(t+s) - Y(t)\|_\infty}{\sigma^*(h)(2\log(1/h))^{1/2}} \ge \sqrt{1+\varepsilon} \right\}$$
$$\le c_\varepsilon \frac{1}{h} \exp\left(-\frac{2}{2+\varepsilon}(1+\varepsilon)\log(1/h) \right) \le c_\varepsilon h^\varepsilon,$$

where $h > 0$ is small enough. From (2.9) of [3], there exists a constant $d > 0$ such that

(2.4)
$$\sigma^{*2}(h) \ge dh^2.$$

For any $\theta > 1$, we define:

$$A_i = \{h : \theta^{-i-1} \le \sigma^*(h) < \theta^{-i}\}, \quad i = 1, 2, \cdots,$$
$$A_{ij} = \{h : \theta^{-j-1} \le h < \theta^{-j}, \ h \in A_i\}, \quad j = 1, 2, \cdots,$$
$$h_{ij} = \sup\{h : h \in A_{ij}\}.$$

Then

(2.5)
$$\limsup_{h \to 0} \sup_{0 \le t \le 1} \sup_{0 \le s \le h} \frac{\|Y(t+s) - Y(t)\|_\infty}{\sigma^*(h)(2\log(1/h))^{1/2}}$$
$$\le \limsup_{i \to \infty} \sup_{j \ge 1} \sup_{0 \le t \le 1} \sup_{0 \le s \le h_{ij}} \frac{\|Y(t+s) - Y(t)\|_\infty}{\theta^{-i-1}(2\log\theta^j)^{1/2}}$$
$$\le \limsup_{i \to \infty} \sup_{j \ge 1} \sup_{0 \le t \le 1} \sup_{0 \le s \le h_{ij}} \frac{\theta^2\|Y(t+s) - Y(t)\|_\infty}{\sigma^*(h_{ij})(2\log(1/h_{ij}))^{1/2}}.$$

By (2.3) and (2.4), we have

$$\sum_{i=1}^\infty P\left\{ \sup_{j \ge 1} \sup_{0 \le t \le 1} \sup_{0 \le s \le h_{ij}} \frac{\|Y(t+s) - Y(t)\|_\infty}{\sigma^*(h_{ij})(2\log(1/h_{ij}))^{1/2}} \ge \sqrt{1+\varepsilon} \right\}$$
$$\le \sum_{i=1}^\infty \sum_{j=1}^\infty P\left\{ \sup_{0 \le t \le 1} \sup_{0 \le s \le h_{ij}} \frac{\|Y(t+s) - Y(t)\|_\infty}{\sigma^*(h_{ij})(2\log(1/h_{ij}))^{1/2}} \ge \sqrt{1+\varepsilon} \right\}$$
$$\le c_\varepsilon \sum_{i=1}^\infty \sum_{j=1}^\infty (\theta^{-j})^{\varepsilon/2}\left(\frac{1}{\sqrt{d}}\sigma^*(h_{ij}) \right)^{\varepsilon/2}$$
$$\le c_\varepsilon \sum_{i=1}^\infty \sum_{j=1}^\infty \theta^{-\varepsilon j/2}(\theta^{\varepsilon/2})^{-i}$$
$$< \infty.$$

The Borel Cantelli lemma yields (1.2) from (2.5) since $\theta > 1$ is arbitrary.

The following Lemma 2.3 is a modification of Corollary 4.2.4 in [6]:

Lemma 2.3. *For each $j = 1, 2, \cdots, n$, let $\{\xi_{kj}\}_{k=1}^{\infty}$ be a sequence of independent jointly standardized normal random variables with $\Lambda_{kj}^{kj'} = \mathrm{Corr}(\xi_{kj}, \xi_{kj'})$ for $j' = 1, 2, \cdots, n$ such that*

$$\delta := \max_{\substack{j \neq j' \\ k \geq 1}} |\Lambda_{kj}^{kj'}| < 1.$$

Let the subsequence $\{l_j\}$ of intergers $\{j : j = 1, \cdots, n\}$ be such that

$$1 \leq l_1 < \cdots < l_m \leq n \quad \text{with} \quad m \leq n.$$

Denote $\lambda_{kj}^{kj'} = \Lambda_{kl_j}^{kl_{j'}}$. Then, for any real $u > 0$,

$$P\left\{ \max_{1 \leq j \leq m} \sup_{k \geq 1} \xi_{kl_j} < u \right\}$$

$$(2.6) \qquad \leq \prod_{k=1}^{\infty} (1 - \Phi(u))^m + c \sum_{\substack{1 \leq j' < j \leq m \\ k \geq 1}} |\lambda_{kj}^{kj'}| \exp\left(\frac{-u^2}{1 + |\lambda_{kj}^{kj'}|} \right),$$

where $c = c(\delta)$ is a constant independent of k, m and u, and we denote

$$\Phi(u) = \int_u^{\infty} \frac{1}{\sqrt{2\pi}} \exp(-y^2/2)\, dy.$$

Estimating an upper bound for the second term of the right hand side of (2.6), we have the following lemma, whose proof is similar to those of [1] and [5]:

Lemma 2.4. *For sufficiently large $M > 0$ and small $h > 0$, let $\xi_{kj} = \{X_k((Mj + 1)h) - X_k(Mjh)\}/\sigma_k(h)$, $k = 1, 2, \cdots$, $j = 1, 2, \cdots, m$, in Lemma 2.3, and put $m = [1/Mh]$, where $[x]$ denotes the greatest integer not exceeding x. Suppose that conditions (ii) and (iii) of Theorem 1.2 hold. Set*

$$u = \frac{\sigma^*(h)}{\sigma_k(h)} \left\{ (2 - \eta) \log \frac{1}{hy_h} \right\}^{1/2},$$

where $0 < \eta < (1 - \delta)\nu/(1 + \nu + \delta)$ for some $\nu > 0$. Then we have

$$\sum_m := \sum_{\substack{1 \leq j' < j \leq m \\ k \geq 1}} |\lambda_{kj}^{kj'}| \exp\left(\frac{-u^2}{1 + |\lambda_{kj}^{kj'}|} \right) \leq ch^{\delta_0},$$

where $\delta_0 = \{\nu(1-\delta) - \eta(1+\nu+\delta)\}/\{(1+\nu)(1+\delta)\} > 0$ and c is a positive constant independent of m, u and k.

Proof. Without loss of generality, let $r = j - j' > 0$ for j, $j' = 1, 2, \cdots, [1/Mh]$. Using the relation $ab = \frac{1}{2}(a^2 + b^2 - (a-b)^2)$, we have

$$
\begin{aligned}
\left|\lambda_{kj}^{kj'}\right| &:= \left|\mathrm{Corr}(\xi_{kj}, \xi_{kj'})\right| \\
&= \frac{1}{\sigma_k^2(h)}\left|E\big[\{(X_k((Mj+1)h) - X_k(Mjh)\} \right. \\
&\qquad \left. \times \{X_k((Mj'+1)h) - X_k(Mj'h)\}\big]\right| \\
&= \frac{1}{2\sigma_k^2(h)}\left|\{E(X_k((Mj+1)h) - X_k(Mj'h))^2 \right. \\
&\qquad + E(X_k(Mjh) - X_k((Mj'+1)h))^2 \\
&\qquad - E(X_k((Mj+1)h) - X_k((Mj'+1)h))^2 \\
&\qquad \left. - E(X_k(Mjh) - X_k(Mj'h))^2\}\right| \\
&= \frac{1}{2\sigma_k^2(h)}\left\{\left|\sigma_k^2((Mj - Mj' + 1)h) + \sigma_k^2((Mj - Mj' - 1)h)\right. \right. \\
&\qquad \left.\left. - \sigma_k^2((Mj - Mj')h) - \sigma_k^2((Mj - Mj')h)\right|\right\} \\
&= \frac{1}{2\sigma_k^2(h)}\left|\int_{Mrh}^{(Mr+1)h}\frac{d\sigma_k^2(x)}{dx}dx - \int_{(Mr-1)h}^{Mrh}\frac{d\sigma_k^2(x)}{dx}dx\right| \\
&\leq \frac{1}{2\sigma_k^2(h)}\int_{(Mr-1)h}^{Mrh}\left(\int_x^{x+h}\left|\frac{d^2\sigma_k^2(y)}{dy^2}\right|dy\right)dx \\
&\leq \frac{1}{2\sigma_k^2(h)}\int_{(Mr-1)h}^{Mrh}\left(\int_x^{x+h}c\frac{\sigma_k^2(y)}{y^2}dy\right)dx \\
&\leq \frac{c}{2\sigma_k^2(h)}\int_{(Mr-1)h}^{Mrh}h\frac{\sigma_k^2(x+h)}{x^2}dx \leq c(Mr)^{2\alpha_k - 2} < r^{-\nu},
\end{aligned}
$$

where $\nu = 1 - \alpha_k > 0$. Let $0 < a = (1-\delta+\eta\delta)/\{(1+\nu)(1+\delta)\} < 1$. We split $\sum_m$ into two sums

$$
\sum_m = \sum_{\substack{1 \leq j' < j \leq m \\ j - j' \leq [m^a] \\ k \geq 1}} |\lambda_{kj}^{kj'}|\exp\left(\frac{-u^2}{1 + |\lambda_{kj}^{kj'}|}\right) + \sum_{\substack{1 \leq j' < j \leq m \\ j - j' > [m^a] \\ k \geq 1}} |\lambda_{kj}^{kj'}|\exp\left(\frac{-u^2}{1 + |\lambda_{kj}^{kj'}|}\right)
$$

$$
=: \sum\nolimits^{(1)} + \sum\nolimits^{(2)}.
$$

92   S. Y. KANG AND Y. K. CHOI

Let us estimate each upper bound of the two sums:

$$\sum{}^{(1)} \le m^{1+a} \sum_{k=1}^{\infty} (hy_h)^{\frac{2-\eta}{1+\delta}\frac{\sigma^{*2}(h)}{\sigma_k^2(h)}}$$

$$\le m^{1+a} \left\{ \sum_{k=1}^{\infty} (hy_h)^{\frac{\sigma^{*2}(h)}{\sigma_k^2(h)}} \right\}^{\frac{2-\eta}{1+\delta}}$$

$$\le c\, h^{\frac{2-\eta}{1+\delta}-1-a} = c\, h^{\delta_0}.$$

Since σ_k are decreasing on $k \ge 1$, it follows from (ii) that

$$\sum{}^{(2)} \le m^{2-a\nu} \sum_{k=1}^{\infty} \exp\left(-u^2 + u^2 m^{-a\nu} \right)$$

$$\le c\, h^{a\nu-2} \sum_{k=1}^{\infty} (hy_h)^{(2-\eta)\sigma^{*2}(h)/\sigma_k^2(h)}$$

$$\le c\, h^{a\nu-2} \left(\sum_{k=1}^{\infty} (hy_h)^{\sigma^{*2}(h)/\sigma_k^2(h)} \right)^{2-\eta}$$

$$= c\, h^{a\nu-\eta} = c\, h^{\delta_0}.$$

Proof of Theorem 1.2. We first prove (1.4). It suffices to prove that, for any $h_n \downarrow 0$ as $n \to \infty$ and $0 < \varepsilon < 1$,

$$\lim_{n\to\infty} P\left\{ \sup_{0\le t\le 1} \frac{\|Y(t+h_n) - Y(t)\|_\infty}{\sigma^*(h_n)(2\log(1/h_n y_{h_n}))^{1/2}} \ge \sqrt{1-\varepsilon} \right\} = 1.$$

Let $m = [1/Mh_n]$, $u = \frac{\sigma^*(h_n)}{\sigma_k(h_n)}\left\{ (2-\eta)\log\frac{1}{h_n y_{h_n}} \right\}^{1/2}$ and $\eta = 2\varepsilon$ in Lemma 2.3. Using (ii) and Lemmas 2.3 and 2.4, we obtain

$$P\left\{ \sup_{0\le t\le 1} \frac{\|Y(t+h_n) - Y(t)\|_\infty}{\sigma^*(h_n)(2\log(1/h_n y_{h_n}))^{1/2}} < \sqrt{1-\varepsilon} \right\}$$

$$\le P\left\{ \max_{1\le j\le [1/Mh_n]} \frac{\|Y((Mj+1)h_n) - Y(Mjh_n)\|_\infty}{\sigma^*(h_n)(2\log(1/h_n y_{h_n}))^{1/2}} < \sqrt{1-\varepsilon} \right\}$$

$$\le P\left\{ \max_{1\le j\le m} \sup_{k\ge 1} \frac{X_k((Mj+1)h_n) - X_k(Mjh_n)}{\sigma_k(h_n)} < u \right\}$$

$$\le \prod_{k=1}^{\infty} (1-\Phi(u))^m + c \sum_{\substack{1\le j'<j\le m \\ k\ge 1}} |\lambda_{kj}^{kj'}| \exp\left(\frac{-u^2}{1+|\lambda_{kj}^{kj'}|} \right)$$

$$\leq \prod_{k=1}^{\infty} \exp\left(-\frac{c}{h_n}(h_n y_{h_n})^{(1-\varepsilon)\frac{\sigma^{*2}(h_n)}{\sigma_k^2(h_n)}}\right) + c\,h_n^{\delta_0}$$

$$\leq \exp\left(-\frac{c}{h_n}\sum_{k=1}^{\infty}(h_n y_{h_n})^{(1-\varepsilon)\frac{\sigma^{*2}(h_n)}{\sigma_k^2(h_n)}}\right) + c\,h_n^{\delta_0}$$

$$\leq \exp\left(-\frac{c}{h_n}\sum_{k=1}^{\infty}(h_n y_{h_n})^{\frac{\sigma^{*2}(h_n)}{\sigma_k^2(h_n)}}(h_n y_{h_n})^{-\varepsilon}\right) + c\,h_n^{\delta_0}$$

$$\leq \exp(-ch_n^{-\varepsilon}) + c\,h_n^{\delta_0} \longrightarrow 0 \quad \text{as} \quad n \to \infty.$$

Hence (1.4) is proved.

Next we prove (1.3). For $\theta > 1$, set

$$B_i = \{h : \theta^{-i-1} \leq \sigma^*(h) < \theta^{-i}\}, \quad i = 1, 2, \cdots,$$
$$B_{ij} = \{h : \theta^{-j-1} \leq hy_h < \theta^{-j}, h \in B_i\}, \quad j = 1, 2, \cdots,$$
$$h'_{ij} = \inf\{h : h \in B_{ij}\}.$$

Then we have

$$\liminf_{h\to 0} \sup_{0\leq t\leq 1} \sup_{0\leq s\leq h} \frac{\|Y(t+s) - Y(t)\|_\infty}{\sigma^*(h)(2\log(1/hy_h))^{1/2}}$$

$$(2.7) \qquad \geq \liminf_{i\to\infty} \sup_{j\geq 1} \sup_{0\leq t\leq 1} \sup_{0\leq s\leq h'_{ij}} \frac{\|Y(t+s) - Y(t)\|_\infty}{\theta^{-i}(2\log\theta^{j+1})^{1/2}}$$

$$\geq \liminf_{i\to\infty} \inf_{j\geq 1} \max_{1\leq l\leq[1/Mh'_{ij}]} \frac{\|Y((Ml+1)h'_{ij}) - Y(Mlh'_{ij})\|_\infty}{\theta^2\sigma^*(h'_{ij})(2\log(1/h'_{ij}y_{h'_{ij}}))^{1/2}}.$$

Let $m = [1/Mh'_{ij}]$ and $u = \frac{\sigma^*(h'_{ij})}{\sigma_k(h'_{ij})}\{(2-\eta)\log\frac{1}{h'_{ij}y_{h'_{ij}}}\}^{1/2}$, $\eta = 2\varepsilon$, in Lemma 2.3. Using Lemmas 2.3 and 2.4, we obtain

$$P\left\{\max_{1\leq l\leq[1/Mh'_{ij}]} \frac{\|Y((Ml+1)h'_{ij}) - Y(Mlh'_{ij})\|_\infty}{\sigma^*(h'_{ij})(2\log(1/h'_{ij}y_{h'_{ij}}))^{1/2}} \leq \sqrt{1-\varepsilon}\right\}$$

$$\leq P\left\{\max_{1\leq l\leq m} \sup_{k\geq 1} \frac{X_k((Ml+1)h'_{ij}) - X_k(Mlh'_{ij})}{\sigma_k(h'_{ij})} \leq u\right\}$$

$$\leq \prod_{k=1}^{\infty}(1-\Phi(u))^m + c\sum_{\substack{1\leq l'<l\leq m \\ k\geq 1}} |\lambda_{kl}^{kl'}|\exp\left(\frac{-u^2}{1+|\lambda_{kl}^{kl'}|}\right)$$

$$=: H_1 + c\,H_2.$$

Estimating an upper bound for H_1, we have, from (2.4),

$$H_1 \leq \exp\left(-c_1 \frac{1}{h'_{ij}} \sum_{k=1}^{\infty} (h'_{ij} y_{h'_{ij}})^{\sigma^{*2}(h'_{ij})/\sigma_k^2(h'_{ij})} (h'_{ij} y_{h'_{ij}})^{-\varepsilon}\right)$$

$$\leq \exp\left(-c_1 (h'_{ij} y_{h'_{ij}})^{-\varepsilon/2} \theta^{i\varepsilon/2})\right)$$

$$\leq \exp\left(-c_1 \theta^{j\varepsilon/2} \theta^{i\varepsilon/2}\right).$$

Next, we estimate an upper bound for H_2. Let $0 < a = (1-\delta+\eta\delta)/\{(1+\nu)(1+\delta)\} < 1$. Then we can split H_2 into two sums:

$$H_2 = \sum_{\substack{1 \leq l' < l \leq m \\ l-l' \leq [m^a] \\ k \geq 1}} |\lambda_{kl}^{kl'}| \exp\left(\frac{-u^2}{1+|\lambda_{kl}^{kl'}|}\right) + \sum_{\substack{1 \leq l' < l \leq m \\ l-l' > [m^a] \\ k \geq 1}} |\lambda_{kl}^{kl'}| \exp\left(\frac{-u^2}{1+|\lambda_{kl}^{kl'}|}\right)$$

$$=: H_3 + H_4.$$

By the same way as in the proof of Lemma 2.4, we can obtain, from (2.4),

$$H_3 \leq \left(\frac{1}{Mh'_{ij}}\right)^{1+a} \sum_{k=1}^{\infty} (h'_{ij} y_{h'_{ij}})^{\frac{2-\eta}{1+\delta}\left(\sigma^{*2}(h'_{ij})/\sigma_k^2(h'_{ij})\right)}$$

$$\leq c\left(\frac{1}{h'_{ij}}\right)^{1+a} \sum_{k=1}^{\infty} (h'_{ij} y_{h'_{ij}})^{\sigma^{*2}(h'_{ij})/\sigma_k^2(h'_{ij})} (h'_{ij} y_{h'_{ij}})^{\left(\frac{2-\eta}{1+\delta}-1\right)\left(\sigma^{*2}(h'_{ij})/\sigma_k^2(h'_{ij})\right)}$$

$$\leq c\left(\frac{1}{h'_{ij}}\right)^{a} (h'_{ij} y_{h'_{ij}})^{\frac{2-\eta}{1+\delta}-1} \leq c(h'_{ij} y_{h'_{ij}})^{\frac{2-\eta}{1+\delta}-1-a}$$

$$= c(h'_{ij} y_{h'_{ij}})^{\delta_0} \leq c(h'_{ij} y_{h'_{ij}})^{\delta_0/2} (\theta^{\delta_0/2})^{-i}$$

$$\leq c\theta^{-\delta_0 j/2} (\theta^{\delta_0/2})^{-i}$$

and

$$H_4 \leq c\left(\frac{1}{h'_{ij}}\right)^{2-a\nu} \sum_{k=1}^{\infty} (h'_{ij} y_{h'_{ij}})^{(2-\eta)\left(\sigma^{*2}(h'_{ij})/\sigma_k^2(h'_{ij})\right)}$$

$$\leq c\left(\frac{1}{h'_{ij}}\right)^{2-a\nu} \sum_{k=1}^{\infty} (h'_{ij} y_{h'_{ij}})^{\sigma^{*2}(h'_{ij})/\sigma_k^2(h'_{ij})} (h'_{ij} y_{h'_{ij}})^{(1-\eta)\left(\sigma^{*2}(h'_{ij})/\sigma_k^2(h'_{ij})\right)}$$

$$\leq c\left(\frac{1}{h'_{ij}}\right)^{1-a\nu} (h'_{ij} y_{h'_{ij}})^{1-\eta} \leq c(h'_{ij} y_{h'_{ij}})^{\delta_0}$$

$$\leq c\theta^{-\delta_0 j/2} \theta^{-\delta_0/2}.$$

So, we get

$$H_1 + cH_2 \le c\theta^{-\delta_0 j/2}\theta^{-\delta_0 i/2}$$

and

$$\sum_{i=1}^{\infty} P\left\{ \inf_{j\ge 1} \max_{1\le l\le [1/Mh'_{ij}]} \frac{\|Y((Ml+1)h'_{ij}) - Y(Mlh'_{ij})\|_{\infty}}{\sigma^*(h'_{ij})(2\log(1/h'_{ij}y_{h'_{ij}}))^{1/2}} \le \sqrt{1-\varepsilon}\right\}$$

$$\le \sum_{i=1}^{\infty}\sum_{j=1}^{\infty} P\left\{ \max_{1\le l\le [1/Mh'_{ij}]} \frac{\|Y((Ml+1)h'_{ij}) - Y(Mlh'_{ij})\|_{\infty}}{\sigma^*(h'_{ij})(2\log(1/h'_{ij}y_{h'_{ij}}))^{1/2}} \le \sqrt{1-\varepsilon}\right\}$$

$$\le c\sum_{i=1}^{\infty}\sum_{j=1}^{\infty} \theta^{-\delta_0 j/2}\theta^{-\delta_0 i/2} < \infty.$$

Using the Borel-Cantelli Lemma, we obtain (1.3) from (2.7) since ε and θ are arbitrary.

References

1. Y. K. Choi, *Erdös-Rényi type laws applied to Gaussian processes*, J. Math. Kyoto Univ. **31** (1991), 191–217.
2. Y. K. Choi and S. H. Lee, *Some limit theorems on the increments of l^p-valued Gaussian processes*, Submitted to Stochastic Analysis and Applications (Nova Science Publishers, Inc.), 2001.
3. M. Csörgő, Z. Y. Lin and Q. M. Shao, *Path properties for l^{∞}-valued Gaussian processes*, Proc. Amer. Math. Soc. **121** (1994), 225–236.
4. X. Fernique, *Continuité des processus Gaussiens*, C. R. Acad. Sci. Paris, t. **258** (1964), 6058-6060.
5. K. S. Hwang, Y. K. Choi and J. S. Kwon, *Moduli of continuity for the increments of a multi-parameter Gaussian processes*, J. Korean Math. **36(4)** (1999), 679–697.
6. M. R. Leadbetter, G. Lindgren and H. Rootzen, *Extremes and Related Properties of Random Sequences and Processes*, Springer-Verlag, New York, 1983.
7. Z. Y. Lin and Y. C. Qin, *On the increments of l^{∞}-valued Gaussian processes*, Preprint.

ON A RANDOM FUNCTIONAL CENTRAL LIMIT THEOREM FOR A STATIONARY MULTIVARIATE LINEAR PROCESS GENERATED BY LINEARLY POSITIVE QUADRANT DEPENDENT RANDOM VECTORS

TAE-SUNG KIM[1] AND MI-HWA KO[1]

ABSTRACT. For a stationary multivariate linear process of the form

$$\mathbb{X}_t = \sum_{j=0}^{\infty} A_j \mathbb{Z}_{t-j},$$

where $\{\mathbb{Z}_t : t = 0, \pm 1, \pm 2, \cdots\}$ is a sequence of stationary linearly positive quadrant dependent m-dimensional random vectors with mean $\mathbb{O}$ and positive definite covariance matrix Γ, we prove a random functional central limit theorem.

1. Introduction

Lehmman [8] introduced a simple and natural definition of positive dependence : A sequence $\{Y_t : t = 0, 1, 2, \cdots\}$ of random variables is said to be *pairwise positive quadrant dependent* (pairwise PQD) if, for any real α_i, α_j and $i \neq j$,

$$P\{Y_i > \alpha_i, Y_j > \alpha_j\} \geq P\{Y_i > \alpha_i\}P\{Y_j > \alpha_j\}.$$

A concept stronger than PQD was introduced by Newman (1984): A sequence $\{Y_t\}$ of random variables is said to be linearly positive quadrant dependent (LPQD) if for any disjoint A, B and positive $r'_j s$ $\sum_{i \in A} r_i Y_i$ and $\sum_{j \in B} r_j Y_j$ are PQD.

Two m-variate random vectors $\mathbb{Z}_1, \mathbb{Z}_2$ are said to be *positive quadrant dependent* (PQD) if $\mathbb{Z}_{1i}, \mathbb{Z}_{2j}$ are PQD for all $i, j = 1, \cdots, m$.

Received December 12, 2000.

1991 *Mathematics Subject Classification*. 60F17, 60G10.

Key Words and Phrases. Multivariate linear process, linearly positive quadrant dependent random vectors, functional central limit theorem.

[1] Division of Mathematical Science, WonKwang University, Iksan, Chonbuk 570-749, Korea. Email: starkim@wonnms.wonkwang.ac.kr

Let $(\mathbb{Z}_1, \mathbb{Z}_2, \cdots, \mathbb{Z}_t)$ be an m-variate random vector. We say that $(\mathbb{Z}_1, \mathbb{Z}_2, \cdots, \mathbb{Z}_t)$ is *linearly positive quadrant dependent* if, for any disjoint $A, B \subset \{1, \cdots, t\}$ and matrix $a_r = (a_{rij})$ with nonnegative components, $i, j = 1, 2, \cdots, m$,

$$(1) \qquad \sum_{s \in A} a_s \mathbb{Z}_s \text{ and } \sum_{r \in B} a_r \mathbb{Z}_r \text{ are PQD.}$$

Let $\{\mathbb{X}_t, t = 0, \pm 1, \cdots\}$ be an m-variate linear process of the form

$$(2) \qquad \mathbb{X}_t = \sum_{u=0}^{\infty} A_u \mathbb{Z}_{t-u}$$

defined on a probability space $(\Omega, \mathcal{F}, P)$, where $\{\mathbb{Z}_t\}$ is a sequence of stationary m-variate LPQD random vectors with $E\mathbb{Z}_t = \mathbb{O}$, $E\|\mathbb{Z}_t\|^2 < \infty$ and positive definite covariance matrix $\Gamma : m \times m$. Throughout this paper, we shall assume that

$$(3) \qquad \sum_{u=0}^{\infty} \|A_u\| < \infty \text{ and } \sum_{u=0}^{\infty} A_u \neq \mathbb{O}_{m \times m},$$

where, for any $m \times m$, $m \geq 1$, matrix $A = (a_{ij})$, $\|A\| = \sum_{i=1}^{m} \sum_{j=1}^{m} |a_{ij}|$ and $\mathbb{O}_{m \times m}$ denotes the $m \times m$ zero matrix. Let W^m denote Wiener process on $C^m[0,1]$, the space of all continuous functions f defined on $[0, 1]$ into $\mathbb{R}^m$ equipped with the norm $\|f\|_\infty = \max_{1 \leq i \leq m} \sup_{0 \leq t \leq 1} |f_i(t)|$.

Further, let

$$T = \left(\sum_{j=0}^{\infty} A_j \right) \Gamma \left(\sum_{j=0}^{\infty} A_j \right)',$$

where the prime denotes transpose, and the matrix $\Gamma = [\sigma_{kj}]$ with $\sigma_{kj} = E(\mathbb{Z}_{1k}\mathbb{Z}_{1j} + \sum_{t=2}^{\infty}(E(\mathbb{Z}_{1k}\mathbb{Z}_{tj}) + E(\mathbb{Z}_{1j}\mathbb{Z}_{tk}))$. Let $\mathbb{S}_n = \sum_{t=1}^{n} \mathbb{X}_t$, $(n \geq 0)(\mathbb{S}_0 = \mathbb{O})$, and define for $n \geq 1$ the stochastic process ξ_n by

$$(4) \qquad \xi_n(u) = n^{-\frac{1}{2}} T^{-\frac{1}{2}} [\mathbb{S}_r + (nu - r)\mathbb{X}_{r+1}], \quad r \leq nu < r + 1,$$

where $r = 0, 1, \cdots, n-1$. In this note we prove a random functional central limit theorem for an m-variate linear process generated by m-variate LPQD random vectors.

Theorem 1.1. *Let* $\{\mathbb{Z}_t, t = 0, \pm 1, \cdots\}$ *be a strictly stationary LPQD sequence of m-dimensional random vectors with* $E(\mathbb{Z}_t) = \mathbb{O}$, $E\|\mathbb{Z}_t\|^2 < \infty$ *and positive definite covariance matrix* Γ *as in (4) and let* $\{\xi_n\}$ *be as in (5). Let*

$\{N_n, n \geq 1\}$ be a sequence of positive integer valued random variables defined on the probability space $(\Omega, \mathcal{F}, P)$ such that, as $n \to \infty$, $N_n/n \xrightarrow{P} N$ with $P(0 < N < \infty) = 1$. Assume

$$(5) \qquad E\|Z_1\|^2 + 2 \sum_{t=2}^{\infty} \sum_{i=1}^{m} E(Z_{1i} Z_{ti}) = \sigma^2 < \infty,$$

$$(6) \qquad \sum_{t=n+1}^{\infty} E\|Z_{1i} Z_{ti}\| = o(n^{-\rho}) \quad \text{for some } \rho > 0,$$

and

$$(7) \qquad E\|Z_t\|^s < \infty \quad \text{for some } s > 2.$$

Then, as $n \to \infty$,

$$\xi_{N_n} \Rightarrow W^m,$$

where $\Rightarrow$ indicates weak convergence, and W^m denotes Wiener proess on $C^m[0, 1]$ the space of all continuous functions f defined on $[0, 1]$ into $\mathbb{R}^m$ equipped with the norm $\|f\|_\infty = \max_{1 \leq i \leq m} \sup_{0 \leq t \leq 1} |f_i(t)|$.

Remark. For $m = 1$, Kim and Baek [7] showed that the process ξ_n defined in (5) converges weakly to W^1, Wiener process on $[0, 1]$.

2. Proofs

Lemma 2.1. Let $\tilde{\mathbb{X}}_t = (\sum_{j=0}^{\infty} A_j) Z_t$ and $\tilde{\mathbb{S}}_k = \sum_{t=1}^{k} \tilde{\mathbb{X}}_k$ and let $\{Z_t\}$ be as in Theorem 1.1. Assume that (6), (7) and (8) hold and that p_n is a sequence of positive integers such that

$$(8) \qquad p_n \to \infty \quad \text{and} \quad p_n/n \to 0 \quad \text{as } n \to \infty.$$

Then

$$(9) \qquad n^{-\frac{1}{2}} \max_{1 \leq k \leq n} \|\tilde{\mathbb{S}}_k - \mathbb{S}_k\| = o_p(1).$$

Proof. See Appendix.

Theorem 2.2. *Let $\{\mathbb{Z}_t\}$ be a sequence of strictly stationary LPQD m-variate random vectors with $E(\mathbb{Z}_t) = \mathbb{O}$, $E\|\mathbb{Z}_t\|^2 < \infty$ and positive definite covariance matrix Γ as in (4) and let $\{\xi_n\}$ be as in (5). If (6), (7) and (8) hold then, as $n \to \infty$,*

$$\xi_n \implies W^m. \tag{10}$$

Proof. Let $\tilde{\xi}_n$ be the same as ξ_n defined in (5) with $\tilde{S}_r$ and $\tilde{\mathbb{X}}_{r+1}$ in place of S_r and $\mathbb{X}_{r+1}$, respectively. By Cramer-Wold device it follows from the multivariate version of Theorem 2.2 of Kim and Baek [7] that $\tilde{\xi}_n \Rightarrow W^m$. Applying Lemma 2.1, we obtain that $\xi_n \Rightarrow W^m$ by Theoren 4.1 of Billingsley [1].

Lemma 2.3. *Let $\{\mathbb{Z}_t\}$ be defined in Theorem 1.1 and let $\tilde{\mathbb{X}}_n$ and $\tilde{\mathbb{S}}_r$ be defined in Lemma 2.1. Assume that (6), (7), (8) and (9) hold and define*

$$\xi_{n,p_n}(u) = \begin{cases} 0, & un \le p_n, \\ n^{-\frac{1}{2}}T^{-\frac{1}{2}}(S_r - S_{p_n}), & p_n \le r \le un < r+1. \end{cases} \tag{11}$$

Then

$$\sup_{0 \le u \le 1} \|\xi_n(u) - \xi_{n,p_n}(u)\| = o_p(1). \tag{12}$$

Proof. Let us first observe that

$$\begin{aligned} \sup_{0 \le u \le 1} & \|\xi_n(u) - \xi_{n,p_n}(u)\| \\ & \le 3n^{-\frac{1}{2}}\|T\|^{-\frac{1}{2}} \max_{1 \le i \le p_n} \|\mathbb{S}_i\| + n^{-\frac{1}{2}}\|T\|^{-\frac{1}{2}} \max_{1 \le i \le n} \|\mathbb{X}_i\|. \end{aligned} \tag{13}$$

By Lemma 2.1, we have

$$n^{-\frac{1}{2}} \max_{1 \le i \le p_n} \|\mathbb{S}_i\| \le n^{-\frac{1}{2}} \max_{1 \le i \le p_n} \|\tilde{\mathbb{S}}_i\| + o_p(1). \tag{14}$$

By Lemma 2 in Kim and Baek [7] and (9),

$$\begin{aligned} n^{-\frac{1}{2}} \max_{1 \le i \le p_n} \|\tilde{\mathbb{S}}_i\| &= o_p(p_n/n) \\ &= o_p(1) \end{aligned} \tag{15}$$

and, by assumptions stationarity and (8),

$$n^{-\frac{1}{2}} \max_{1 \le i \le n} \|\mathbb{X}_i\| = o_p(1). \tag{16}$$

Combining these observations (14), (15) and (16) shows that the right-hand side of (13) converges in probability to zero and hence the proof of Lemma 2.3 is complete.

Proof of Theorem 1.1. First note that $\xi_{n,p_n} \Rightarrow W^m$, which follows directly from an application of Theorem 2.2, Lemma 2.3 and Theorem 4.1 of Billingsley [1]. Combining this result and Lemma 2.3 we arrive at (17.19) of Billingsley [1]. From this point, the proof of Theorem 1.1 follows along the same lines as that given in his Theorem 17.2 and hence the details are omitted.

Appendix

Because Lemma 2.1 plays the important role to prove the main result, we prove Lemma 2.1 by using the ideas in the proof of Lemma 3 of [3] and Lemma 2 of [7].

Proof of Lemma 2.1. As in the proof of Lemma 3 of [5], we observe that

$$\tilde{S}_k = \sum_{t=1}^{k}\left(\sum_{j=0}^{k-t} A_j\right) \mathbb{Z}_t + \sum_{t=1}^{k}\left(\sum_{j=k-t+1}^{\infty} A_j\right) \mathbb{Z}_t$$

$$= \sum_{t=1}^{k}\left(\sum_{j=0}^{t-1} A_j \mathbb{Z}_{t-j}\right) + \sum_{t=1}^{k}\left(\sum_{j=k-t+1}^{\infty} A_j\right) \mathbb{Z}_t$$

and so

$$\tilde{S}_k - S_k = -\sum_{t=1}^{k}\sum_{j=t}^{\infty} A_j \mathbb{Z}_{t-j} + \sum_{t=1}^{k}\left(\sum_{j=k-t+1}^{\infty} A_j\right) \mathbb{Z}_t$$

$$= I_1 + I_2 \text{ (say)}.$$

To prove

$$(A.1) \qquad\qquad n^{-\frac{1}{2}} \max_{1 \le k \le n} \|I_1\| \xrightarrow{P} 0,$$

102   T. S. KIM AND M. H. KO

we observe that, for $r > 2$,

$$n^{-\frac{r}{2}} E \max_{1 \leq k \leq n} \left\| \sum_{t=1}^{k} \sum_{j=t}^{\infty} A_j \mathbb{Z}_{t-j} \right\|^r$$

$$= n^{-\frac{r}{2}} E \max_{1 \leq k \leq n} \left\| \sum_{j=1}^{\infty} \sum_{t=1}^{j \wedge k} A_j \mathbb{Z}_{t-j} \right\|^r$$

$$\leq n^{-\frac{r}{2}} \left(\sum_{j=1}^{\infty} \|A_j\| \left\{ E \max_{1 \leq k \leq n} \left\| \sum_{t=1}^{j \wedge k} \mathbb{Z}_{t-j} \right\|^r \right\}^{\frac{1}{r}} \right)^r$$

$$\leq K \left[\sum_{j=1}^{\infty} \|A_j\| \left(\frac{j \wedge k}{n} \right)^{\frac{1}{2}} \right]^r$$

for some positive constant K by using Lemma 2 in Kim and Baek [7] for LPQD random variables. By the dominated convergence theorem, the last term above tends to zero as $n \to \infty$ and then we prove (A.1).

Next, we show that

(A.2) $$\qquad\qquad n^{-\frac{1}{2}} \max_{1 \leq k \leq n} \|I_2\| = o_p(1).$$

Write

$$I_2 = I_{21} + I_{22},$$

where

$$I_{21} = A_1 \mathbb{Z}_k + A_2 (\mathbb{Z}_k + \mathbb{Z}_{k-1}) + \cdots + A_k (\mathbb{Z}_k + \cdots + \mathbb{Z}_1),$$
$$I_{22} = (A_{k+1} + A_{k+2} + \cdots)(\mathbb{Z}_k + \cdots + \mathbb{Z}_1).$$

Note that

$$n^{-\frac{1}{2}} \max_{1 \leq k \leq n} \|I_{22}\| \leq \left(\sum_{i=0}^{\infty} \|A_i\| \right) n^{-\frac{1}{2}} \max_{1 \leq k \leq p_n} \|\mathbb{Z}_1 + \cdots + \mathbb{Z}_k\|$$

$$+ \left(\sum_{i > p_n} \|A_i\| \right) n^{-\frac{1}{2}} \max_{1 \leq k \leq n} \|\mathbb{Z}_1 + \cdots + \mathbb{Z}_k\|$$

$$= III + IV \text{ (say)}.$$

It follows from (3) and (9) that, for some $r > 2$,

$$III \leq \left(\sum_{i=0}^{\infty} \|A_i\| \right)^r B_1 (p_n/n)^{\frac{r}{2}} \xrightarrow{P} 0$$

and

$$IV \leq \left(\sum_{i>p_n} \|A_i\| \right)^r B_2 \xrightarrow{P} 0.$$

By Lemma 2 of Kim and Baek [7] and it remains to prove that

$$Y_n := n^{-\frac{1}{2}} \max_{1 \leq k \leq n} \|I_{21}\| = o_p(1).$$

To this end, define, for each $l \geq 1$,

$$I_{21,l} = B_1 \mathbb{Z}_k + B_2(\mathbb{Z}_k + \mathbb{Z}_{k+1}) + \cdots + B_k(\mathbb{Z}_k + \cdots + \mathbb{Z}_1),$$

where

$$B_k = \begin{cases} A_k, & k \leq l, \\ \mathbb{O}_{m \times m}, & k > l. \end{cases}$$

Let $Y_{n,l} = n^{-\frac{1}{2}} \max_{1 \leq k \leq n} \|I_{21,l}\|$. Clearly, for each $l \geq 1$,

(A.3)
$$Y_{n,l} = o_p(1).$$

On the other hand, it follows that

$$(Y_{n,l} - Y_n)$$

$$\leq n^{-\frac{1}{2}} \max_{1 \leq k \leq n} \left\| \sum_{i=1}^{k} (A_i - B_i)\mathbb{Z}_k + \cdots + \mathbb{Z}_{k-i+1}) \right\|$$

$$\leq n^{-\frac{1}{2}} \max_{l < k \leq n} \left(\sum_{i=l+1}^{k} \|A_i\| \max_{l < i \leq n} \|\mathbb{Z}_k + \cdots + \mathbb{Z}_{k-i+1}\| \right)$$

$$\leq n^{-\frac{1}{2}} \sum_{i>l} \|A_i\| \max_{1 \leq k \leq n} \max_{l < i \leq k} \left(\|\mathbb{Z}_1 + \cdots + \mathbb{Z}_k\| + \|\mathbb{Z}_1 + \cdots + \mathbb{Z}_{k-i}\| \right)$$

$$\leq n^{-\frac{1}{2}} \sum_{i>l} \|A_i\| \left(\max_{l < k \leq n} \|\mathbb{Z}_1 + \cdots + \mathbb{Z}_k\| + \max_{l < k \leq n} \max_{l < i \leq n} \|\mathbb{Z}_1 + \cdots + \mathbb{Z}_{k-i}\| \right)$$

$$\leq n^{-\frac{1}{2}} \sum_{i>l} \|A_i\| \left(\max_{1 \leq j \leq n} \|\mathbb{Z}_1 + \cdots + \mathbb{Z}_j\| + \max_{1 \leq k \leq n} \|\mathbb{Z}_1 + \cdots + \mathbb{Z}_j\| \right)$$

$$= 2n^{-\frac{1}{2}} \sum_{i>l} \|A_i\| \max_{1 \leq j \leq n} \|\mathbb{Z}_1 + \cdots + \mathbb{Z}_j\|.$$

From this result and Lemma 2 of [7], for any $\delta > 0$,

$$\lim_{l \to \infty} \lim_{n \to \infty} \sup P(|Y_{n,l} - Y_n|^2 > \delta)$$

(A.4)
$$\leq \lim_{l \to \infty} 2^r \delta^{-r} \left(\sum_{i > l} \|A_i\| \right)^r \lim_{n \to \infty} n^{-\frac{r}{2}} \max_{1 \leq j \leq n} \|\mathbb{Z}_1 + \cdots + \mathbb{Z}_j\|^r$$

$$\leq B \lim_{l \to \infty} \delta^{-r} 2^r \left(\sum_{i > l} \|A_i\| \right)^r$$

$$= 0.$$

In view of (A.3) and (A.4), it follows from Theorem 4.2 of Billingsley [1] that $Y_n = o_p(1)$. This completes the proof of Lemma 2.1.

REFERENCES

1. P. Billingsley, *Convergence of Probability Measures*, Wiley, New York, 1968.
2. P. Birkel, *A functional central limit theorem for positively dependent random variables*, J. Multi. Anal. **44** (1993), 314–320.
3. J. Esary, F. Proschan, and D. Walkup, *Association of random variables with applications*, Ann. Math. Statist. **38** (1967), 1466–1474.
4. I. Fakhre-Zakeri and S. Lee, *Sequential estimation of the mean vector of a multivariate linear process*, J. Multivariate Anal. **47** (1993), 196–209.
5. ______, *On functional central limit theorems for multivariate linear process with applications to sequential estimation*, J. Stat. Planning and Inference **83** (2000), 11–23.
6. A. Gut, *Stopped Random Walks, Limit Theorems and Applications*, Springer, New York, 1988.
7. T. S. Kim and J. I. Baek, *A central limit theorem for the stationary linear processes generated by linearly positive quadrant dependent processes*, Stat. and Probab. Letts.(accepted) (2000).
8. E. L. Lehmann, *Some concepts of dependence*, Ann. Math. Statist. **37** (1966), 1137–1153.

MARKOV CHAIN ADAPTIVE DESIGNS

ZHENGYEN LIN[1]

ABSTRACT. The best possible medical treatment for the individual patient. Various adaptive designs have beIn any clinical trial on a cohort of human beings, there is an ethical imperative to provide en proposed towards this goal. Since Zelen's famous play-the-winner rule, there are a lot of discussion about it and its generalization, randomized play-the-winner rule, in the literature. In this paper, we propose a family of adaptive designs, which can be viewed as another generalization of Zelen's play-the-winner rule. We called it Markov chain adaptive design. Some asymptotic properties and advantages of the Markov chain adaptive design are studied.

1. Introduction

Traditional designs for clinical trials use the balanced (or 50-50) allocation of patients to treatments. But, from an ethical view of point, one may want to reduce the total number of failure outcomes in a trial and keep the capability of making a comparison.

Adaptive designs, an important subdivision of experimental designs, are designs in which the probability a treatment is assigned to the coming patient depends upon the results of the previous patients in the study. The goal is to skew assignment probabilities to favor the treatment performing better thus far in the study.

Consider a two-arm clinical trial: Two treatments (A and B) with dichotomous response (success and failure). Patients are recruited into the clinical trial sequentially. Zelen (1969) propose the following design, so-called "Play Winner Rule" (PWR): *A success on a particular treatment generates a future trial on the*

Received October 27, 2000

1991 *Mathematiccs Subject Classification.* Primary 92B05, 60J20, 62P10; Secondary 60j10.

Key Words and Phrases. Adaptive designs, asymptotic normality, play-the-winner rule, randomized play-the-winner rule

Project supported by NSFC (19571021) and NSFZP (195043).

[1] Department of Mathematics, Zhejiang University, Hangzhou, Zhejiang 310028, Peoples Republic of China. Email: zlin@public1.hz.zj.cn

same treatment with a new patient. A failure on a treatment generates a future trial on the alternate treatment.

Motivated as an extension to Zelen's (1969) idea, Wei and Durham (1978) introduced the following randomized play-the-winner rule: We start with (γ, γ) balls (types A and B, respectively) in the urn. If a type A (resp., B) ball is drawn, the patient is assigned to treatment A (resp., B). The ball is replaced and the patient response is observed. A success on treatment A (resp., B) or a failure on treatment B (resp., A) generates a type A (resp., B) ball in the urn.

One large family of randomized adaptive designs can be developed from the generalized Polya urn (GPU) model. Smythe (1996) defines the Extended Polya Urn model (EPU) (a special class of GPU model) and considers its asymptotic normality. Bai and Hu (1999) establish consistency and asymptotic normality for GPU model with nonhomogeneous generating matrix. Wei (1979) first noted that the randomized play-the-winner rule could be formulated as a Generalized Polya Urn model.

In this paper, we propose a new class of adaptive design, we called it Markov chain adaptive designs, which is another extension to Zelen's (1969) play-the-winner rule.

2. Markov Chain Adaptive Design

Suppose that, at stage i, the treatment A is assigned to the patient i. Then the $(i+1)$th patient will be assigned either treatment A or treatment B according certain probabilities, which depend on the response of the ith patient. Let α_s be the probability to assign the $(i+1)$th patient to treatment A, when the response of the ith patient is "success" with treatment A. Let α_f be the probability to assign the $(i+1)$th patient to treatment B, when the response of the ith patient is "failure" with treatment A. Similarly, let β_s be the probability to assign the $(i+1)$th patient to treatment B, if the i th patient is with treatment B and "success". And β_f is the probability to assign the $(i+1)$th patient to treatment A, if the i th patient is with treatment B and "failure".

When $\alpha_s = 1$, $\alpha_f = 0$, $\beta_s = 1$ and $\beta_f = 0$, we get Zelen's play-the-winner. When $\alpha_s = \alpha_f$ and $\beta_s = \beta_f$, we get another special case, which does not depend on any previous information of the trial. Let

$$\begin{cases} X_i = 1, & \text{treatment A is assigned to } i\text{th patient,} \\ 0, & \text{the treatment B is assigned to } i\text{th patient.} \end{cases}$$

Then $S_n = \sum_{i=1}^{n} X_i$ is the number of the treatment A in the first n stages. Now let $p_{iA} = P(\text{"success"}|X_i = 1)$ and $p_{iB} = P(\text{"success"}|X_i = 0)$. When $p_{iA} = p_A$ and $p_{iB} = p_B$ for all i, $\{X_i\}$ is a homogeneous Markov chain and

$$\begin{pmatrix} \alpha & 1-\beta \\ 1-\alpha & \beta \end{pmatrix},$$

where $\alpha = p_A\alpha_s + (1-p_A)\alpha_f$ and $\beta = p_B\beta_s + (1-p_B)\beta_f$. In general, if p_{iA} and p_{iB} depend on i, $\{X_i\}$ is a nonhomogeneous Markov chain. The ith transition matrix is

$$\begin{pmatrix} \alpha_i & 1-\beta_i \\ 1-\alpha_i & \beta_i \end{pmatrix},$$

where $\alpha_i = p_{iA}\alpha_s + (1-p_{iA})\alpha_f$ and $\beta_i = p_{iB}\beta_s + (1-p_{iB})\beta_f$.

3. Asymptotic Properties

At first, we consider a homogeneous Markov chain with the transition matrix is

$$\begin{pmatrix} \alpha & 1-\beta \\ 1-\alpha & \beta \end{pmatrix}.$$

That is $P(X_{i+1} = 1 | X_i = 1) = \alpha$ and $P(X_{i+1} = 0 | X_i = 0) = \beta$.

Theorem 1. *For a homogeneous Markov chain X_i with transition*

$$\begin{pmatrix} \alpha & 1-\beta \\ 1-\alpha & \beta \end{pmatrix},$$

we have

$$\frac{S_n}{n} \to^P \frac{1-\beta}{2-\alpha-\beta}$$

and

$$\sqrt{n}(S_n/n - \frac{(1-\beta)}{2-\alpha-\beta}) \to^D N(0,\sigma^2) \quad \text{as } n \to \infty,$$

where

$$\sigma^2 = \frac{(1-\alpha)(1-\beta)(\alpha+\beta)}{(2-\alpha-\beta)^3}.$$

When patients exhibit a drift in characteristics over time, we need to consider the nonhomogeneous Markov chain. Assume that the transition matrix is

$$\begin{pmatrix} \alpha_i & 1-\beta_i \\ 1-\alpha_i & \beta_i \end{pmatrix}.$$

Furthermore, assume that there exist α and β with $0 < \alpha < 1$ and $0 < \beta < 1$ such that $\alpha_i \to \alpha$ and $\beta_i \to \beta$ as $i \to \infty$.

Theorem 2. *For the nonhomogeneous Markov chain X_i, suppose that*

$$d_i := \max\{|\alpha_i - \alpha|, |\beta_i - \beta|\} \to 0.$$

Then the first conclusion of Theorem 1 remains true. If it is further assumed that $\sqrt{i}d_i \to 0$, then the second conclusion of Theorem 1 is true.

4. Discussion

For the randomized play-the-winner rule, let N_n represent the number of patients with treatment A in the first n patients. Athreya and Karlin (1968) showed that

$$\frac{N_n}{n} \to \frac{q_B}{q_A + q_B} \quad \text{almost surely as } n \to \infty.$$

When $p_A + p_B < 1.5$, we have (cf., Smythe and Rosenberger (1995))

$$\sqrt{n}\left(\frac{N_n}{n} - \frac{q_B}{q_A + q_B}\right) \to N(0, \sigma_1^2)$$

in distribution, where

$$\sigma_1^2 = \frac{[5 - 2(q_A + q_B)]q_A q_B}{[2(q_A + q_B) - 1](q_A + q_B)^2}.$$

When $p_A + p_B > 1.5$, the limiting distributions of the proportion of patients assigned to treatment A are unknown. For the Markov chain adaptive designs, we know that the asymptotic normality of S_n/n is true for any p_A and q_B from Theorem 1. So these designs can be used for $p_A + p_B > 1.5$.

Now we compare the play-the-winner rule with the randomized play-the-winner rule. We can easily see that the proportions of both designs converge to $q_B/(q_A + q_B)$. Because of $\sigma_1^2 > \sigma^2$, we can say that the play-the-winner rule is more stable than the randomized play-the-winner rule, even for the case $p_A + p_B < 1.5$. Further, we can choose the value of α and β to minimize the asymptotic variance of the proportion, but keep the same $q_B/(q_A + q_B)$. Therefore, we get more stable Markov chain adaptive designs.

REFERENCES

1. K. B. Athreya and S. Karlin, *Embedding of urn schemes into continuous time branching processes and related limit theorems*, Ann. Math. Statist. **39** (1968), 1801–1817.
2. Z. D. Bai and F. Hu, *Asymptotic theorems for urn models with nonhomogeneous generating matrix*, Stochastic Process. Appl. **80** (1999), 87–101.

3. R. T. Smythe, *Central limit theorems for urn models.*, Stochastic Process. Appl. **65** (1996), 115–137.

4. R. T. Smythe and W. F. Rosenberger, *Play-the winner designs, generalized Polya urns, and Markov branching processes,*, In Adaptive Designs (N. Flournoy and W. F. Rosenberger, eds.), Hayward, CA: Institute of Mathematical Statistics, pp. 13–22, 1995.

5. L. J. Wei, *The generalized Polya's urn design for sequential medical trials*, Ann. Statist. **7** (1979), 291–296.

6. L. J. Wei and S. Durham, *The randomized play-the-winner rule in medical trials*, J. Amer. Statist. Assoc. **73** (1978), 840–843.

7. M. Zelen, *Play the winner rule and the controlled clinical trial*, J. Amer. Statist. Assoc. **64** (1969), 131–146.

PATH PROPERTIES OF THE
PRIMITIVES OF A BROWNIAN MOTION

ZHENGYAN LIN[1]

ABSTRACT. Let $\{W(t);\ t \geq 0\}$ be a standard Brownian motion. For a positive integer m, define a Gaussian process

$$X_m(t) = \int_0^t (t-s)^m dW(s).$$

In this paper, we study the bounds of its moduli of continuity and large increments by establishing large deriation results.

1. Introduction

Let $\{W(t);\ t \geq 0\}$ be a standard Brownian motion. For a positive integer m, define a Gaussian process

$$(1) \qquad X_m(t) = \frac{1}{m!} \int_0^t (t-s)^m dW(s),$$

which was first mentioned by Shepp [6]. It is easy to see that $X_0(t) = W(t)$ and

$$X_m(t) = \int_0^t X_{m-1}(s)ds, \quad t \geq 0,\ m \geq 1.$$

So we call it the m'th integrated Brownian motion or the m-fold primitive. This class of processes arises in several domains of applied mathematics. For instance, the process $X_1(\cdot)$, which has been studied at length, is the solution of the Langevin's equation under certain physical conditions. Wahba [7], [8] used

Received November 14, 2000.

1991 *Mathematiccs Subject Classification.* Primary 60F15, 60G65, 60G15; Secondary 60G17.

Key Words and Phrases. Moduli of continuity, large increment, Brownian motion, primitive.

Project supported by NSFC (19571021) and NSFZP (195043).

[1] Department of Mathematics, Zhejiang University, Hangzhou, Zhejiang 310028, People's Republic of China. Email: zlin@mail.hz.zj.cn

$X_m(\cdot)$ to derive a correspondence between smoothing by splines and Bayesian estimation in certain stochastic models.

In [9], Watanabe established a law of the iterated logarithm for $X_1(\cdot)$ (in fact, his result concerns a larger class of Gaussian processes). Lachal [3], [4] studied the law of the iterated logarithm and regular points for $X_m(\cdot)$ for $m \geq 1$.

In [5], Lin studied path behavior of the process $X_m(\cdot)$. By establishing results on large deviations, we show the moduli of continuity and large increment properties for $X_m(\cdot)$ for all $m \geq 1$, which are the analogues of the corresponding results for the Brownian motion $W(\cdot)$ (cf. [2]). Note that increments of $X_m(\cdot)$ are neither independent nor stationary.

First of all, we give some moment results. We have

$$
(2) \qquad EX_m^2(t) = \frac{1}{(m!)^2} \int_0^t (t-s)^{2m} ds =: b_m t^{2m+1},
$$

where $b_m = (m!)^{-2}(2m+1)^{-1}$ and, for any $h > 0$,

$$
\begin{aligned}
&E(X_m(t+h) - X_m(t))^2 \\
&= \frac{1}{(m!)^2} E\left(\int_0^{t+h} (t+h-s)^m dW(s) - \int_0^t (t-s)^m dW(s) \right)^2 \\
&= \frac{1}{(m!)^2}\{ E\left(\int_0^t \left(\sum_{j=1}^m \binom{m}{j}(t-s)^{m-j}h^j \right) dW(s) \right)^2 \\
&\qquad + E\left(\int_t^{t+h} (t+h-s)^m dW(s) \right)^2 \} \\
&=: \sum_{j=2}^{2m+1} b_{mj} h^j t^{2m+1-j}
\end{aligned}
$$

(3)

for some positive b_{mj}, $j = 2, ..., 2m+1$, where $b_{m2} = ((m-1)!)^{-2}(2m-1)^{-1}$, which implies that

$$
(4) \qquad E(X_m(t+h) - X_m(t))^2 = (1 + \delta(h/t))b_{m2}h^2 t^{2m-1},
$$

where $0 < \delta(x) \to 0$ as $x \to 0$. Hence we have

$$
\begin{aligned}
&E(X_m(t+h)X_m(t)) \\
&= \frac{1}{2}E\{X_m^2(t+h) + X_m^2(t) - (X_m(t+h) - X_m(t))^2\} \\
&= \frac{1}{2}b_m((t+h)^{2m+1} + t^{2m+1}) - \frac{1}{2}\sum_{j=2}^{2m+1} b_{mj}h^j t^{2m+1-j}.
\end{aligned}
$$

(5)

Put $Y_m(t) = X_m(t)/t^{m-1/2}$. By (2), we have

$$(6) \qquad EY_m^2(t) = b_m t^2.$$

Using (2), (3) and (5), we have

$$
\begin{aligned}
& E(Y_m(t+h) - Y_m(t))^2 \\
&= E\Big\{ \frac{1}{(t+h)^{m-1/2}}(X_m(t+h) - X_m(t)) \\
& \qquad - \Big(\frac{1}{t^{m-1/2}} - \frac{1}{(t+h)^{m-1/2}}\Big) X_m(t) \Big\}^2 \\
&= \frac{1}{(t+h)^{2m-1}} \sum_{j=2}^{2m+1} b_{mj} h^j t^{2m+1-j} \\
& \quad + \frac{((t+h)^{m-1/2} - t^{m-1/2})^2}{t^{2m-1}(t+h)^{2m-1}} b_m t^{2m+1} - \frac{2((t+h)^{m-1/2} - t^{m-1/2})}{t^{m-1/2}(t+h)^{2m-1}} \\
& \qquad \times \Big\{ \frac{1}{2} b_m((t+h)^{2m+1} - t^{2m+1}) - \frac{1}{2}\sum_{j=2}^{2m+1} b_{mj} h^j t^{2m+1-j} \Big\} \\
&=: B_m h^2 + g_m(h,t),
\end{aligned}
$$

(7)

where

$$
\begin{aligned}
B_m &= b_{m2} + b_m\Big\{ (m - \tfrac{1}{2})^2 - (m - \tfrac{1}{2})(2m+1) \Big\} \\
&= b_{m2} - b_m\Big(m - \frac{1}{2}\Big)\Big(m + \frac{3}{2}\Big),
\end{aligned}
$$

$$g_m(h,t) = O(h^3 t)$$

as $ht \to 0$, which implies that

$$(8) \qquad E(Y_m(t+h) - Y_m(t))^2 = (1 + o(1))B_m h^2$$

as $ht \to 0$.

2. Large Deviations

The following is a large deviation result for small time increments:

114   Z. LIN

Theorem 2.1. *For any $\varepsilon > 0$, there exist positive numbers h_0, x_0, c_1 and C_1 such that, for any $0 < h \leq h_0$ and $x \geq x_0$,*

$$
(9) \qquad P\left(\sup_{0<t\leq 1-h} \sup_{0\leq s\leq h} \frac{|X_m(t+s) - X_m(t)|}{(t \vee h)^{m-1/2}} \geq (1+\varepsilon)b_{m2}^{1/2}hx \right)
$$
$$
\leq C_1(e^{-c_1 x^2} + h^{-1}e^{-x^2/2}),
$$

where $a \vee b = \max\{a, b\}$.

An analogue of Theorem 2.1 in the large increment case is the following:

Theorem 2.2. *Let a_T be a function of T with $0 < a_T \leq T$ and $a_T/T \to 0$ as $T \to \infty$. Then, for any $\varepsilon > 0$, there exist positive numbers T_0, x_1, c_2 and C_2 such that, for any $T \geq T_0$ and $x \geq x_1$,*

$$
(10) \qquad P\left(\sup_{0<t\leq T-a_T} \sup_{0\leq s\leq a_T} \frac{|X_m(t+s) - X_m(t)|}{(t \vee a_T)^{m-1/2}} \geq (1+\varepsilon)b_{m2}^{1/2}a_Tx \right)
$$
$$
\leq C_2(e^{-c_2 x^2} + Ta_T^{-1}e^{-x^2/2}).
$$

3. Moduli of Continuity and Large Increments

Theorem 3.1. *We have*

$$
(11) \qquad \lim_{h\to 0} \sup_{0<t\leq 1-h} \sup_{0\leq s\leq h} \frac{|X_m(t+s) - X_m(t)|}{b_{m2}^{1/2}(t \vee h)^{m-1/2}h(2\log h^{-1})^{1/2}} = 1, \quad a.s.,
$$

$$
(12) \qquad \lim_{h\to 0} \sup_{0<t\leq 1-h} \frac{|X_m(t+h) - X_m(t)|}{b_{m2}^{1/2}(t \vee h)^{m-1/2}h(2\log h^{-1})^{1/2}} = 1. \quad a.s.
$$

Theorem 3.2. *Let a_T be a continuous function of T with $0 < a_T \leq T$ and suppose that*

(i) $\displaystyle \lim_{n\to\infty} \sup_{n-1<t\leq n} a_t \Big/ \inf_{n-1<t\leq n} a_t = 1,$

(ii) $\displaystyle \lim_{T\to\infty} \log(T/a_T)/\log\log T = \infty.$

Then

$$
(13) \qquad \lim_{T\to\infty} \sup_{0<t\leq T-a_T} \sup_{0\leq s\leq a_T} \frac{|X_m(t+s) - X_m(t)|}{b_{m2}^{1/2}(t \vee a_T)^{m-1/2}a_T(2\log(T/a_T))^{1/2}}
$$
$$
= 1 \quad a.s.,
$$

$$
(14) \qquad \lim_{T\to\infty} \sup_{0<t\leq T-a_T} \frac{|X_m(t+a_T) - X_m(t)|}{b_{m2}^{1/2}(t \vee a_T)^{m-1/2}a_T(2\log(T/a_T))^{1/2}}
$$
$$
= 1 \quad a.s.
$$

4. Further Results on Large Increments

What happen if the condition (ii) fails? In [10] and [11], Wang studied this question and showed the following:

Theorem 4.1. *Let a_T be a function of T with $0 < a_T \leq T$ and suppose that*
(iii) *a_T and T/a_T are monotonically non-decreasing,*
(iv) $\lim_{T\to\infty} \log(T/a_T)/\log\log T = r,\ 0 \leq r \leq \infty.$
Then

(15)
$$\liminf_{T\to\infty}\ \sup_{0<t\leq T-a_T}\ \sup_{0\leq s\leq a_T} \frac{|X_m(t+s) - X_m(t)|}{b_{m2}^{1/2}(t \vee a_T)^{m-1/2}a_T(2(\log(T/a_T) + \log\log T))^{1/2}}$$
$$\leq \alpha(r) \quad \text{a.s.,}$$

where

(16)
$$\alpha(r) = \begin{cases} \sqrt{\dfrac{r}{r+1}}, & 0 \leq r < \infty, \\ 1, & r = \infty. \end{cases}$$

If, instead of the condition (iv), a_T satisfies
(v) $\lim_{T\to\infty} \log\log(T/a_T)/\log\log\log T = r,\ 1 < r \leq \infty,$
then there exists a constant $\gamma > 0$ which satisfies that, for any $0 < \varepsilon < 2$ and $0 < \delta < r - 1$,

$$\gamma \begin{cases} < \sqrt{\dfrac{(2-\varepsilon)(r-\delta-1)(r+1)}{2r(r+1-\delta)}} & \text{if } 1 < r < \infty, \\ = 1 & \text{if } r = \infty, \end{cases}$$

such that

(17)
$$\liminf_{T\to\infty}\ \sup_{0<t\leq T-a_T} \frac{|X_m(t+a_T) - X_m(t)|}{b_{m2}^{1/2}(t \vee a_T)^{m-1/2}a_T(2(\log\log(T/a_T) + \log\log\log T))^{1/2}}$$
$$\geq \gamma\alpha(r) \quad \text{a.s.}$$

Theorem 4.2. *Let a_T be a function of T with $0 < a_T \leq T$ and suppose that*
(vi) $\lim_{T\to\infty} \log(T/a_T)/\log\log\log T = \infty.$

Then

$$
\liminf_{T\to\infty}\ \sup_{0<t\le T-a_T}\ \sup_{0\le s\le a_T}
$$

$$
(18)\qquad \frac{|X_m(t+s)-X_m(t)|}{b_{m2}^{1/2}(t\vee a_T)^{m-1/2}a_T(2(\log(T/a_T)-\log\log\log T))^{1/2}}
$$

$$
\le 1\quad a.s.
$$

Investigating Theorem 4.1, one can know that (17) is not available for the case of $\lim\limits_{T\to\infty}\log\log(T/a_T)/\log\log\log T\le 1$. This motivates to introduce the following problem:

Find the normalizing factor instead of that of (17).

Actually, in [10], Wang showed the following result: Put

$$
\Phi(x)=\frac{1}{\sqrt{2\pi}}\int_{-\infty}^{x}e^{-t^2/2}dt.
$$

Theorem 4.3. *Let a_T be a function of T with $0<a_T\le T$ and suppose that*
(vii) $\lim\limits_{T\to\infty}(T/a_T)/(\log T)^{A\log\log\log T}=0$ *for* $A=(2+\delta)/(-\log\Phi(1))$, $0<\delta<1$.
Then

$$
(19)\qquad \liminf_{T\to\infty}\ \sup_{0<t\le T-a_T}\ \sup_{0\le s\le a_T}\frac{|X_m(t+s)-X_m(t)|}{b_{m2}^{1/2}t^{m-1/2}a_T\left(\log\left(1+\frac{T}{a_T(\log T)^{A\log\log\log T}}\right)\right)}
$$

$$
\ge 1\quad a.s.
$$

5. Laws of Iterated Logarithm

Denote $C_0[0,1]=(C_0[0,1],\|\cdot\|_\infty)$ by the set $C_0[0,1]$ of all continuous functions f on $[0,1]$ with $f(0)=0$, endowed with the sup-norm $\|f\|_\infty=\sup\limits_{0\le t\le 1}|f(t)|$, and denote by $H_\mu\subset C_0[0,1]$ the reproducing kernel Hilbert space corresponding to the Gaussian measure $\mu=\mathcal{L}(X_m)$ on the seperable Banach space $C_0[0,1]$. If $f\in H_\mu$, then $\|f\|_\mu$ denotes the H_μ–norm of f. Put

$$
\varphi_m(t)=t^{m+1/2}\sqrt{2b_m\log\log\frac{1}{t}},
$$

$$
\psi_m(t)=t^{m+1/2}\sqrt{2b_m\log\log t},\quad t>0.
$$

In [3], Lachel showed the following laws of the iterated logarithm:

Theorem 5.1. *We have*

$$(20) \qquad \limsup_{T \to 0^+} \sup_{0 \le s \le T} \frac{|X_m(s)|}{\varphi_m(T)} = 1 \quad a.s.,$$

$$(21) \qquad \limsup_{t \to \infty} \sup_{0 \le s \le T} \frac{|X_m(s)|}{\psi_m(T)} = 1 \quad a.s.$$

In [10], Wang showed the following functional laws of the iterated logarithm:
Set $K_m = \{f \in H_\mu; \|f\|_\mu \le \frac{1}{\sqrt{b_m}}\}$.

Theorem 5.2. *We have*

$$(22) \qquad \lim_{T \to 0^+} \inf_{f \in K_m} \left\| \frac{X_m(T \cdot)}{\varphi_m(T)} - f(\cdot) \right\|_\infty = 0 \quad a.s.$$

and, for each $f \in K_m$,

$$(23) \qquad \liminf_{T \to 0^+} \left\| \frac{X_m(T \cdot)}{\varphi_m(T)} - f(\cdot) \right\|_\infty = 0 \quad a.s.$$

Theorem 5.3. *We have*

$$(24) \qquad \lim_{T \to \infty} \inf_{f \in K_m} \left\| \frac{X_m(T \cdot)}{\psi_m(T)} - f(\cdot) \right\|_\infty = 0 \quad a.s.$$

and, for each $f \in K_m$,

$$(25) \qquad \liminf_{T \to \infty} \left\| \frac{X_m(T \cdot)}{\psi_m(T)} - f(\cdot) \right\|_\infty = 0 \quad a.s.$$

For Chung's law of iterated logarithm, Chen and Li [1] showed the following result:

Theorem 5.4. *There exists constant $\kappa_m > 0$ which depends only on m such that*

$$(26) \qquad \liminf_{T \to \infty} \left(\frac{\log \log T}{T} \right)^{m+1/2} \sup_{0 \le s \le T} |X_m(s)| = \kappa_m \quad a.s.$$

REFERENCES

1. X. Chen and W. V. Li, *Quadratic functionals and small ball probabilities for the m-fold integrated Brownian motion*, Preprint (1999).
2. M. Csörgő and P. Révész, *Strong Approximations in Probability and Statistics*, Academic Press, New York-Akadémiai Kiadó, Budapest, 1981.
3. A. Lachel, *Local asymptotic classes for the successive primitives of Brownian motion*, Ann. Probab 25 (1997), 1712–1734.
4. A. Lachal, *Regular points for the successive primitives of Brownian motion*, to appear in J. Kyodo Math. Univ (1997).
5. Z. Lin, *The moduli of continuity and large increments of the primitives of a Brownian motion*, Preprint (1999).
6. L. A. Shepp, *Radon-Nikodym derivatives of Gaussian measures*, Ann. Math. Statist. 37 (1966), 321–354.
7. G. Wahba, *Improper priors, spline smoothing and the problem of guarding against model error in regression*, J. Roy. Statist. Soc. Ser. B 40 (1978), 364–372.
8. G. Wahba, *Bayesian "confidence intervals" for the cross-validated smoothing spline*, J. Roy. Statist. Soc. Ser. B 45 (1983), 133–150.
9. H. Watanabe, *An asymptotic property of Gaussian processes I*, Trans. Amer. Math. Soc. 148 (1970), 233–248.
10. W. Wang, *On some liminf results of the increments of the primitive of Brownian motion*, Preprint (2000).
11. W. Wang, *Functional law of the iterated logarithm of the m-fold integrated Brownian motion*, Preprint (2000).

SHARP STRONG LIMIT THEOREMS ON THE INCREMENTS OF A GAUSSIAN PROCESS

HEE JIN MOON[1] AND YONG KAB CHOI[1]

ABSTRACT. Some sharp strong limit theorems on the increments of a Gaussian process are established by estimating upper bounds of large deviation probabilities on the suprema of the Gaussian process. Our results generalize some theorems in [4] and [1].

1. Introduction and Results

Several strong limit theorems on the increments of Wiener processes, Gaussian processes and partial sum processes have been studied by many authors, for instance, Csörgő and Révész [5], Shao [11], Lin and Lu [10], Kono [8]-[9], Choi [1], Choi and Kono [2], Csörgő and Steinebach [6], Deheuvels and Steinebach [7], etc.

In this paper, we are interested in investigating the almost surely exact convergence for the difference between suprema on the increments of a Gaussian process and a normalizing factor. Throughout the paper, we shall assume the following statements: Let $\{X(t), 0 \leq t < \infty\}$ be an almost surely continuous, centered Gaussian process on the probability space $(\Omega, \mathcal{S}, P)$ with $X(0) = 0$ and stationary increments $E\{X(t) - X(s)\} = \sigma^2(|t - s|)$, where $\sigma(\cdot)$ is an increasing continuous and regularly varying function on $(0, \infty)$ with exponent α at 0 and ∞ for some $0 < \alpha < 1$. Futher, assume that there exists a positive constant c such that, for $t > 0$,

$$\left| \frac{d^2 \sigma^2(t)}{dt^2} \right| \leq c \frac{\sigma^2(t)}{t^2}.$$

Received November 9, 2000.

1991 *Mathematics Subject Classification.* 60F15, 60G15.

Key Words and Phrases. Regularly varying function, Gaussian process.

This work was partially supported by Korea Research Foundation Grant (KRF-99-005-D00003).

[1] Department of Mathematics, Gyeongsang National University, Chinju 660-701, Korea. Email: mhj0307@orgio.net and mathykc@nongae.gsnu.ac.kr

Let $a_T \, (0 < T < \infty)$ be a real-valued function of T satisfying the following conditions:

    (i) $0 < a_T \leq T$,
    (ii) a_T is non-decreasing,
    (iii) $\liminf_{T \to 0} a_T > 0$,
    (iv) $\lim_{T \to \infty} \log(T/a_T) / \log \log T = \infty$.

For convenience, we denote:

$$\gamma_T = \{2(\log(T/a_T) + \log \log T)\}^{1/2}, \quad \log x = \ln(x \vee 1),$$
$$\delta_T = \sigma^{-1}(\sigma(a_T)/\gamma_T) \vee 1,$$
$$d_T = \log((a_T/\delta_T) \vee 1) - \log \gamma_T + \varepsilon \log \log T \quad \text{for any small } \varepsilon > 0,$$
$$u_T = \gamma_T + (d_T/\gamma_T), \quad x_T = \sigma(a_T)u_T,$$

where $\sigma^{-1}(\cdot)$ denotes the inverse function of $\sigma(\cdot)$ and $a \vee b = \max\{a, b\}$.

The main results are as follows:

Theorem 1.1. *Let $\delta_T = 1$. Then we have*

$$(1.1) \quad \lim_{T \to \infty} \frac{\gamma_T}{\sigma(a_T)d_T} \left(\sup_{0 \leq t \leq T} \sup_{0 \leq s \leq a_T} |X(t+s) - X(t)| - \sigma(a_T)\gamma_T \right) = 1 \quad \text{a.s.}$$

Theorem 1.2. *Let $\delta_T > 1$. Then we have*

$$(1.2) \quad \lim_{T \to \infty} \frac{\sigma(a_T)d_T}{\gamma_T} \left(\sup_{0 \leq t \leq T} \sup_{0 \leq s \leq a_T} |X(t+s) - X(t)| - \sigma(a_T)\gamma_T \right) = 1 \quad \text{a.s.}$$

As we will be seen in Lemma 2.1, note that the range of a_T is changed according as $\delta_T = 1$ or $\delta_T > 1$. For two functions f and g, we define: $f(T) \sim g(T)$ if $\lim_{T \to \infty} f(T)/g(T) = 1$. We know that if $\delta_T = 1$, then $\sigma(a_T)d_T/\gamma_T \to 0$ as $T \to \infty$ from Lemma 2.1(1) and we have

$$\sup_{0 \leq t \leq T} \sup_{0 \leq s \leq a_T} |X(t+s) - X(t)| - \sigma(a_T)\gamma_T \sim \sigma(a_T)d_T/\gamma_T$$

from Theorem 1.1. This means that the speed converging to zero for the difference between $\sup_{0 \leq t \leq T} \sup_{0 \leq s \leq a_T} |X(t+s) - X(t)|$ and a normalizing factor $\sigma(a_T)\gamma_T$ is exactly the same as that of $\sigma(a_T)d_T/\gamma_T$ converging to zero. Similar arguments are stated also in Theorem 1.2: If $\delta_T > 1$, then $\gamma_T/(\sigma(a_T)d_T) \to 0$ as $T \to \infty$ from Lemma 2.1(2) and we have

$$\sup_{0 \leq t \leq T} \sup_{0 \leq s \leq a_T} |X(t+s) - X(t)| - \sigma(a_T)\gamma_T \sim \gamma_T/(\sigma(a_T)d_T)$$

from Theorem 1.2. We remark that Theorems 1.1 and 1.2 are much general considering the following two corollaries:

Corollary 1.1. *We have*

$$\lim_{T\to\infty}\left(\sup_{0\le t\le T}\sup_{0\le s\le a_T}|X(t+s)-X(t)|-\sigma(a_T)\gamma_T\right)=0\quad\text{a.s.}$$

Corollary 1.1 implies the following Corollary 1.2. But, it is clear that Corollary 1.1 and Theorems 1.1 and 1.2 cannot be in general obtained from Corollary 1.2.

Corollary 1.2. *([1], [3], [4]) We have*

$$\lim_{T\to\infty}\sup_{0\le t\le T}\sup_{0\le s\le a_T}\frac{|X(t+s)-X(t)|}{\sigma(a_T)\gamma_T}=1\quad\text{a.s.}$$

2. The Proofs

The following Lemmas 2.1~2.2 are essential to prove the theorems:

Lemma 2.1. *Let $0<\varepsilon<1/2$. For large $T>0$, we have the following:*

(1) *If $\delta_T=1$, then $a_T\le(2\log T)^{\frac{1}{2}-\varepsilon}$, $0>d_T\downarrow-\infty$, $\frac{d_T}{\gamma_T}\uparrow 0$ and $\frac{\gamma_T}{\sigma(a_T)d_T}\downarrow$ $-\infty$ as $T\uparrow\infty$.*

(2) *If $\delta_T>1$, then $a_T>(2\log T)^{\frac{1}{2}-\varepsilon}$, $0<d_T\uparrow\infty$ and $\frac{\sigma(a_T)d_T}{\gamma_T}\uparrow\infty$ as $T\uparrow\infty$.*

(3) *$x_T\uparrow\infty$ as $T\uparrow\infty$.*

Proof. (1) Suppose that $\delta_T=1$. Then $\sigma(a_T)\le\gamma_T$. Since $\sigma(t)=t^\alpha L(t)$, $t>0$, where $L(\cdot)$ is a slowly varying function, we have

$$\sigma(a_T)\le\gamma_T\implies a_T^{2\alpha}L(a_T)^2+2\log a_T\le 2(\log T+\log\log T)$$
$$\implies a_T^{2\alpha}L(a_T)^2\le a_T^{2/(1-2\varepsilon)}\le 2\log T$$
$$\implies a_T\le(2\log T)^{\frac{1}{2}-\varepsilon}.$$

Thus, for large T,

$$d_T\le\frac{1}{2}\log\log T-\frac{1}{2}\log\left(\log T+\left(\frac{1}{2}+\varepsilon\right)\log\log T-\left(\frac{1}{2}-\varepsilon\right)\log 2\right)$$
$$<0$$

and $d_T\downarrow-\infty$ as $T\uparrow\infty$. Also

$$d_T\ge-\frac{1}{2}\log 2-\frac{1}{2}\log(\log T+\log\log T)+\varepsilon\log\log T.$$

Since $0 < a_T \le (2\log T)^{\frac{1}{2}-\varepsilon}$, we have

$$\left\{2\left(\log T + \left(\frac{1}{2} + \varepsilon\right)\log\log T - \left(\frac{1}{2} - \varepsilon\right)\log 2\right)\right\}^{1/2}$$
$$\le \gamma_T \le \{2(\log T + \log\log T)\}^{1/2}.$$

From these facts, it is easy to see that $d_T/\gamma_T \uparrow 0$, $\gamma_T/(\sigma(a_T)d_T) \downarrow -\infty$ and $x_T \sim \sigma(a_T)\gamma_T \uparrow \infty$ as $T \uparrow \infty$.

(2) If $\delta_T > 1$, then $\sigma(a_T) > \gamma_T$ and $a_T > \sigma^{-1}(\gamma_T)$. Since

$$\gamma_T > 1 \iff \sigma(a_T) > \sigma(a_T)/\gamma_T$$
$$\iff a_T > \delta_T = \sigma^{-1}(\sigma(a_T)/\gamma_T) > 1,$$

we have $a_T > \delta_T > 1$. By the regular variation of $\sigma(\cdot)$, we get

$$\sigma(a_T/\gamma_T)/\sigma(a_T) \ge 1/\gamma_T$$

and hence $a_T/(\delta_T\gamma_T) \ge 1$. Therefore, we have

$$d_T = \log\left(a_T/(\delta_T\gamma_T)\right) + \varepsilon\log\log T \ge \varepsilon\log\log T > 0$$

and $x_T > \sigma(a_T)\gamma_T \uparrow \infty$ as $T \uparrow \infty$. Since $\sigma(a_T) > \gamma_T$, we have

$$\frac{\sigma(a_T)d_T}{\gamma_T} > d_T \uparrow \infty \quad \text{as } T \uparrow \infty$$

and, further,

$$\sigma(a_T) > \gamma_T \implies a_T^2 + 2\log a_T > 2(\log T + \log\log T)$$
$$\implies a_T > (2\log T)^{\frac{1}{2}-\varepsilon}.$$

Lemma 2.2. [1] *For any $\varepsilon > 0$, there exists a positive constant C_ε such that, for all $u > 0$,*

$$P\left\{\sup_{0 \le t \le T} \sup_{0 \le s \le a_T} \frac{|X(t+s) - X(t)|}{\sigma(a_T)} \ge u\right\} \le C_\varepsilon \frac{T}{a_T}\exp\left(-\frac{u^2}{2+\varepsilon}\right).$$

Proof of Theorem 1.1. For any $0 < \varepsilon < 1/2$, let $1 < \theta < \left(\frac{1+2\varepsilon}{1+\varepsilon}\right)^{1/(4(1+\alpha))}$, $0 < \alpha < 1$. For $k \ge 1$, let $T_k \le T \le T_{k+1}$ and put

$$A_{kl} = \{T_k : \theta^{k-1} \le T_k \le \theta^k, \ \theta^{l-1} \le a_{T_k} \le \theta^l\}, \quad -\infty < l < \infty.$$

First we will prove that almost surely there exists $T_0 > 0$ such that, for all $T \geq T_0$,

$$(2.1) \qquad \sup_{0 \leq t \leq T} \sup_{0 \leq s \leq a_T} |X(t+s) - X(t)| - \sigma(a_T)\gamma_T \leq \frac{\sigma(a_T)d_T}{\gamma_T}.$$

Since a_T is nondecreasing, we get $\theta^l \leq a_{T_{k+1}} \leq \theta^{l+1}$ from A_{kl}. Thus, by the regualr variation of $\sigma(\cdot)$, we have

$$(2.2) \qquad \begin{aligned} \frac{\sigma(a_{T_k})}{\sigma(a_{T_{k+1}})} &\geq \frac{\sigma(a_{T_k})\gamma_{T_k}}{\sigma(a_{T_{k+1}})\gamma_{T_{k+1}}} \geq \theta^{-2(1+\alpha)} \\ &> \left(\frac{1+\varepsilon}{1+2\varepsilon}\right)^{1/2} > \left(\frac{1-3\varepsilon}{1-2\varepsilon}\right)^{1/2}. \end{aligned}$$

From (2.2) and by Lemmas 2.1 and 2.2, we have, for any small $\varepsilon > 0$ and large k,

$$\begin{aligned} P&\left\{ \sup_{0 \leq t \leq T_{k+1}} \sup_{0 \leq s \leq a_{T_{k+1}}} |X(t+s) - X(t)| - \sigma(a_{T_k})\gamma_{T_k} > \sqrt{1+2\varepsilon}\,\frac{\sigma(a_{T_k})d_{T_k}}{\gamma_{T_k}} \right\} \\ &\leq P\left\{ \sup_{0 \leq t \leq T_{k+1}} \sup_{0 \leq s \leq a_{T_{k+1}}} |X(t+s) - X(t)| \geq \sqrt{1+2\varepsilon}\,\sigma(a_{T_k})u_{T_k} \right\} \\ &\leq P\left\{ \sup_{0 \leq t \leq T_{k+1}} \sup_{0 \leq s \leq a_{T_{k+1}}} \frac{|X(t+s) - X(t)|}{\sigma(a_{T_{k+1}})} \geq \sqrt{1+\varepsilon}\,u_{T_k} \right\} \\ &\leq C_\varepsilon \frac{T_{k+1}}{a_{T_{k+1}}} \frac{1}{\gamma_{T_k}} \exp\left\{ -\frac{1+(\varepsilon/2)}{2+\varepsilon}(\gamma_{T_k}^2 + 2d_{T_k}) \right\} \\ &\leq C_\varepsilon(\log T_k)^{-1-\varepsilon} \leq C_\varepsilon k^{-1-\varepsilon}. \end{aligned}$$

By the Borel-Cantelli lemma, it follows that almost surely there exists $k_0 > 0$ such that, for all $k \geq k_0$,

$$\sup_{0 \leq t \leq T_{k+1}} \sup_{0 \leq s \leq a_{T_{k+1}}} |X(t+s) - X(t)| - \sigma(a_{T_k})\gamma_{T_k} \leq \frac{\sigma(a_{T_k})d_{T_k}}{\gamma_{T_k}}.$$

This implies (2.1) from Lemma 2.1 (1).

Next we prove that almost surely there exists $T_0 > 0$ such that, for all $T \geq T_0$,

$$(2.3) \qquad \sup_{0 \leq t \leq T} \sup_{0 \leq s \leq a_T} |X(t+s) - X(t)| - \sigma(a_T)\gamma_T \geq \frac{\sigma(a_T)d_T}{\gamma_T}.$$

By (iv), there exists $M > 0$ large enough such that $T_k/a_{T_k} > M$ for all large k and define positive integers h_{T_k} by $h_{T_k} = [T_k/(Ma_{T_k})]$. For $i = 0, 1, \cdots, h_{T_k}$, we also define

$$X_{T_k}(i) = X((Mi+1)a_{T_k}) - X(Mia_{T_k}).$$

Then $X_{T_k}(i)/\sigma(a_{T_k})$ are standard normal random variables. Using (iv), (2.2) and $d_T < 0$, it follows that, for any small $\varepsilon > 0$ and large k,

$$
\begin{aligned}
P\Bigg\{ &\sup_{0 \le t \le T_k} \sup_{0 \le s \le a_{T_k}} |X(t+s) - X(t)| - \sigma(a_{T_{k+1}})\gamma_{T_{k+1}} \\
&\qquad < \sqrt{1 - 3\varepsilon}\, \frac{\sigma(a_{T_{k+1}})d_{T_{k+1}}}{\gamma_{T_{k+1}}} \Bigg\} \\
&\le P\Bigg\{ \sup_{0 \le t \le T_k} \frac{|X(t+a_{T_k}) - X(t)|}{\sigma(a_{T_k})} \le \sqrt{1 - 3\varepsilon}\, \frac{\sigma(a_{T_{k+1}})}{\sigma(a_{T_k})} \left(\gamma_{T_{k+1}} + \frac{d_{T_{k+1}}}{\gamma_{T_{k+1}}} \right) \Bigg\} \\
&\le P\Bigg\{ \sup_{0 \le t \le T_k} \frac{|X(t+a_{T_k}) - X(t)|}{\sigma(a_{T_k})} \le \sqrt{1 - 2\varepsilon}\, \gamma_{T_k} \Bigg\} \\
&\le P\Bigg\{ \max_{0 \le i \le h_{T_k}} \frac{X_{T_k}(i)}{\sigma(a_{T_k})} \le \sqrt{(2 - 2\varepsilon) \log h_{T_k}} \Bigg\}.
\end{aligned}
$$

The remainder of the proof is similar to the corresponding proof of Theorem 2.2 in [1]. Thus it follows that almost surely there exists $k_0 > 0$ such that, for all $k \ge k_0$,

$$\sup_{0 \le t \le T_k} \sup_{0 \le s \le a_{T_k}} |X(t+s) - X(t)| - \sigma(a_{T_{k+1}})\gamma_{T_{k+1}} \ge \frac{\sigma(a_{T_{k+1}})d_{T_{k+1}}}{\gamma_{T_{k+1}}}.$$

This implies (2.3) from Lemma 2.1 (1). Combining (2.1) with (2.3), we obtain that almost surely there exists $T_0 > 0$ such that, for all $T \ge T_0$,

$$\sup_{0 \le t \le T} \sup_{0 \le s \le a_T} |X(t+s) - X(t)| - \sigma(a_T)\gamma_T = \frac{\sigma(a_T)d_T}{\gamma_T}.$$

This completes the proof of Theorem 1.1.

Proof of Theorem 1.2. The proof is similar to that of Theorem 1.1 and we shall still keep the same notations as in that proof. We first prove that almost surely there exists $T_0 > 0$ such that, for all $T \ge T_0$,

$$(2.4) \qquad \sup_{0 \le t \le T} \sup_{0 \le s \le a_T} |X(t+s) - X(t)| - \sigma(a_T)\gamma_T \le \frac{\gamma_T}{\sigma(a_T)d_T}.$$

Again by (2.2), Lemma 2.2 and $d_T > 0$, we have, for any small $\varepsilon > 0$ and large k,

$$
\begin{aligned}
P\bigg\{ & \sup_{0 \le t \le T_{k+1}} \sup_{0 \le s \le a_{T_{k+1}}} |X(t+s) - X(t)| - \sigma(a_{T_k})\gamma_{T_k} \\
& > \sqrt{1+2\varepsilon}\, \frac{\gamma_{T_{k+1}}}{\sigma(a_{T_{k+1}})d_{T_{k+1}}} \bigg\} \\
\le P\bigg\{ & \sup_{0 \le t \le T_{k+1}} \sup_{0 \le s \le a_{T_{k+1}}} |X(t+s) - X(t)| \\
& \ge \sqrt{1+2\varepsilon}\, \sigma(a_{T_{k+1}})\gamma_{T_{k+1}} \bigg(\frac{\sigma(a_{T_k})\gamma_{T_k}}{\sigma(a_{T_{k+1}})\gamma_{T_{k+1}}} + \frac{1}{\sigma^2(a_{T_{k+1}})d_{T_{k+1}}} \bigg) \bigg\} \\
\le P\bigg\{ & \sup_{0 \le t \le T_{k+1}} \sup_{0 \le s \le a_{T_{k+1}}} \frac{|X(t+s) - X(t)|}{\sigma(a_{T_{k+1}})} \ge \sqrt{1+2\varepsilon}\, \gamma_{T_{k+1}} \frac{\sigma(a_{T_k})\gamma_{T_k}}{\sigma(a_{T_{k+1}})\gamma_{T_{k+1}}} \bigg\} \\
\le P\bigg\{ & \sup_{0 \le t \le T_{k+1}} \sup_{0 \le s \le a_{T_{k+1}}} \frac{|X(t+s) - X(t)|}{\sigma(a_{T_{k+1}})} \ge \sqrt{1+\varepsilon}\, \gamma_{T_{k+1}} \bigg\} \\
\le C_\varepsilon & \frac{T_{k+1}}{a_{T_{k+1}}} \exp\bigg(-\frac{1+\varepsilon}{2+\varepsilon}\gamma_{T_{k+1}}^2 \bigg) \le C_\varepsilon k^{-1-\varepsilon'},
\end{aligned}
$$

where $\varepsilon' = \varepsilon/(2+\varepsilon)$. So, the Borel-Cantelli lemma yields that almost surely there exists $k_0 > 0$ such that, for all $k \ge k_0$,

$$
\sup_{0 \le t \le T_{k+1}} \sup_{0 \le s \le a_{T_{k+1}}} |X(t+s) - X(t)| - \sigma(a_{T_k})\gamma_{T_k} \le \frac{\gamma_{T_{k+1}}}{\sigma(a_{T_{k+1}})d_{T_{k+1}}}.
$$

This implies (2.4) from Lemma 2.1 (2).

Next we prove that almost surely there exists $T_0 > 0$ such that, for all $T \ge T_0$,

$$
(2.5) \qquad \sup_{0 \le t \le T} \sup_{0 \le s \le a_T} |X(t+s) - X(t)| - \sigma(a_T)\gamma_T \ge \frac{\gamma_T}{\sigma(a_T)d_T}.
$$

Since $d_{T_k} \to \infty$ as $k \to \infty$ from Lemma 2.1(2), it follows from (2.2) that for large k

$$
\frac{\sigma(a_{T_{k+1}})\gamma_{T_{k+1}}}{\sigma(a_{T_k})\gamma_{T_k}} + \frac{1}{\sigma^2(a_{T_k})\gamma_{T_k}} \le \bigg(\frac{1-2\varepsilon}{1-8\varepsilon} \bigg)^{1/2}.
$$

Thus by (iv) we have, for any small $\varepsilon > 0$ and large k,

$$
P\Bigg\{ \sup_{0\leq t\leq T_k} \sup_{0\leq s\leq a_{T_k}} |X(t+s) - X(t)| - \sigma(a_{T_{k+1}})\gamma_{T_{k+1}}
$$
$$
< \sqrt{1 - 8\varepsilon}\, \frac{\gamma_{T_k}}{\sigma(a_{T_k})d_{T_k}} \Bigg\}
$$
$$
\leq P\Bigg\{ \sup_{0\leq t\leq T_k} \sup_{0\leq s\leq a_{T_k}} |X(t+s) - X(t)|
$$
$$
\leq \sqrt{1 - 8\varepsilon}\,\sigma(a_{T_k})\gamma_{T_k} \left(\frac{\sigma(a_{T_{k+1}})\gamma_{T_{k+1}}}{\sigma(a_{T_k})\gamma_{T_k}} + \frac{1}{\sigma^2(a_{T_k})d_{T_k}} \right) \Bigg\}
$$
$$
\leq P\Bigg\{ \sup_{0\leq t\leq T_k} \frac{|X(t+a_{T_k}) - X(t)|}{\sigma(a_{T_k})} \leq \sqrt{1 - 2\varepsilon}\gamma_{T_k} \Bigg\}
$$
$$
\leq P\Bigg\{ \max_{0\leq i\leq h_{T_k}} \frac{X_{T_k}(i)}{\sigma(a_{T_k})} \leq \left((2 - 2\varepsilon)\log h_{T_k}\right)^{1/2} \Bigg\}.
$$

As in the corresponding proof of Theorem 1.1, it follows that almost surely there exists $k_0 > 0$ such that, for all $k \geq k_0$,

$$
\sup_{0\leq t\leq T_k} \sup_{0\leq s\leq a_{T_k}} |X(t+s) - X(t)| - \sigma(a_{T_{k+1}})\gamma_{T_{k+1}} \geq \frac{\gamma_{T_k}}{\sigma(a_{T_k})d_{T_k}}.
$$

This implies (2.5) from Lemma 2.1 (5). Combining (2.4) with (2.5) yields (1.2).

References

1. Y. K. Choi, *Erdös-Rényi type laws applied to Gaussian processes*, J. Math. Kyoto Univ., **31(1)** (1991), 191–217.
2. Y. K. Choi and N. Kôno, *On the asymptotic behavior of Gaussian sequences with stationary increments*, J. Math. Kyoto Univ., **31(3)** (1991), 643–678.
3. E. Csáki, M. Csörgő, Z. Y. Lin and P. Révész, *On infinite series of independent Ornstein-Uhlenbeck processes*, Stoch. Processes & their Appl., **39** (1991), 25–44.
4. M. Csörgő and P. Révész, *How big are the increments of a Wiener process?* Ann. Probab., **7** (1979), 731–737.
5. M. Csörgő and P. Révész, *Strong Approximations in Probability and Statistics*, Academic Press, New York, 1981.
6. M. Csörgő and J. Steinebach, *Improved Erdös-Rényi and strong approximation laws for increments of partial sums*, Ann. Probab., **9(6)** (1981), 988–996.
7. P. Deheuvels and J. Steinebach, *Exact convergence rates in strong approximation laws for large increments of partial sums*, Probab. Th. Rel. Fields, **76** (1987), 369–393.
8. N. Kôno, *On the asymptotic behavior of the increments of a Wiener process*, J. Math. Kyoto Univ., **31** (1991), 289–306.
9. N. Kôno, *The exact modulus of cntinuity for Gaussian processes taking values of a finite dimensional normed space*, Trends in Probab. & Related Analysis, SAP'96, World Scientific, Singapore, 1996, 219–232.

10. Z. Y. Lin and C. R. Lu, *Strong Limit Theorems*, Kluwer Acad. Publ., New York, 1992.
11. Q. M. Shao, *Remark on increments of the Wiener process*, J. of Math., **6** (1986), 175–182; in Chinese.

C^*-ALGEBRAS OVER SPHERES WITH FIBRES NONCOMMUTATIVE TORI OF RANK 3

CHUN-GIL PARK[1]

ABSTRACT. All C^*-algebras of sections of locally trivial C^*-algebra bundles over $\prod_{i=1}^e S^{2n_i}$ with fibres completely irrational noncommutative tori A_ω of rank 3 are constructed and classified.

0. Introduction

Given a locally compact abelian group G and a multiplier ω on G, one can associate to them the twisted group C^*-algebra $C^*(G, \omega)$, which is the universal object for unitary ω-representations of G. $C^*(\mathbb{Z}^m, \omega)$ is said to be a *noncommutative torus of rank m* and denoted by A_ω. The multiplier ω determines a subgroup S_ω of G, called its *symmetry group*, and the multiplier ω is called *totally skew* if the symmetry group S_ω is trivial. And A_ω is called *completely irrational* if ω is totally skew. See [1, 10, 16]. It was shown in [1] that if G is a locally compact abelian group and ω is a totally skew multiplier on G, then $C^*(G, \omega)$ is a simple C^*-algebra.

The noncommutative torus A_ω is the universal object for unitary ω-representations of $\mathbb{Z}^m$, so A_ω is realized as $C^*(u_1, \cdots, u_m \mid u_i u_j = e^{2\pi i \theta_{ji}} u_j u_i)$, where u_i are unitaries and θ_{ji} are real numbers for $1 \leq i, j \leq m$. Boca [3] showed that almost all completely irrational noncommutative tori are realized as inductive limits of circle algebras, where the term *"circle algebra"* denotes a C^*-algebra which is a finite direct sum of algebras of the form $C(\mathbb{T}^1) \otimes M_q(\mathbb{C})$. In several papers [3, 14], whether the completely irrational noncommutative torus A_ω can be realized as an inductive limit of circle algebras was studied. We will assume

---

Received November 9, 2000.

1991 *Mathematics Subject Classification.* Primary 46L87, 46L05; Secondary 55R15.

Key Words and Phrases. Homogeneous C^*-algebra, C^*-algebra bundle, real rank 0, circle algebra, crossed product.

This work was supported by grant No. 1999-2-102-001-3 from the interdisciplinary Research program year of the KOSEF.

[1] Department of Mathematics, Chungnam National University, Taejon 305-764, Korea
Email: cgpark@math.chungnam.ac.kr

that each completely irrational noncommutative torus appearing in this paper is an inductive limit of circle algebras.

Each d-homogeneous C^*-algebra over M is isomorphic to the C^*-algebra $\Gamma(\eta)$ of sections of a locally trivial C^*-algebra bundle η with base space M, fibre $M_d(\mathbb{C})$, and structure group $\mathrm{Aut}(M_d(\mathbb{C})) \cong PU(d)$. See [13, 15] for details. So each d-homogeneous C^*-algebra over $\prod_{i=1}^e S^{2n_i} \times \mathbb{T}^2$ is realized as the C^*-algebra $\Gamma(\zeta)$ of sections of a locally trivial C^*-algebra bundle ζ over $\prod_{i=1}^e S^{2n_i} \times \mathbb{T}^2$ with fibres $M_d(\mathbb{C})$.

In this paper, we are going to construct all C^*-algebras of sections of locally trivial C^*-algebra bundles over $\prod_{i=1}^e S^{2n_i}$ with fibres completely irrational noncommutative tori A_ω of rank 3, and we are going to classify them.

1. Homogeneous C^*-algebras over a Product Space of Spheres

The set $[M, BPU(d)]$ of homotopy classes of continuous maps of a compact Hausdorff space M into the classifying space $BPU(d)$ of the Lie group $PU(d)$ is in bijective correspondence with the set of equivalence classes of principal $PU(d)$-bundles over M, which is in bijective correspondence with the set of d-homogeneous C^*-algebras over M. The homotopy group $[S^{2n}, BPU(d)]$ classifies d-homogeneous C^*-algebras over S^{2n}. $[S^{2n}, BPU(d)] = [S^{2n-1}, PU(d)] \cong \mathbb{Z}$ if $n > 1$, $\cong \mathbb{Z}_d$ if $n = 1$, which are the cyclic groups. So each group has a generator, and there is a unitary $U(z) \in PU(d)$ such that the generating d-homogeneous C^*-algebra over S^{2n} can be realized as the C^*-algebra of sections of a locally trivial C^*-algebra bundle over S^{2n} characterized by the unitary $U(z) \in PU(d)$ over S^{2n-1}. If $(d, a) = p$ ($|p| > 1$), then consider the d-homogeneous C^*-algebra over S^{2n} corresponding to each $a \in \mathbb{Z}$ or $\mathbb{Z}_d$ as the tensor product of $M_p(\mathbb{C})$ with a $\frac{d}{p}$-homogeneous C^*-algebra over S^{2n}, which is given by $U(z)^{\frac{a}{p}} \in PU(\frac{d}{p})$. Consider $U(z)^a$ as $U(z)^{\frac{a}{p}} \otimes I_p \in PU(d)$, where I_p denotes the $p \times p$ identity matrix. Then every d-homogeneous C^*-algebra $B_{d,a}$ over S^{2n} can be realized as the C^*-algebra of sections of a locally trivial C^*-algebra bundle over S^{2n} characterized by the unitary $U(z)^a \in PU(d)$ over S^{2n-1} for some $a \in \mathbb{Z}$ or $\mathbb{Z}_d$.

Krauss and Lawson [13] proved that every d-homogeneous C^*-algebra over S^{2n} is isomorphic to one of the following C^*-subalgebras $B_{d,a}$, $a \in \mathbb{Z}$ or $\mathbb{Z}_d$, of $C(e_+^{2n} \amalg e_-^{2n}, M_d(\mathbb{C}))$, given as follows: if $f \in B_{d,a}$, then the following condition is satisfied

$$f_+(z) = U(z)^a f_-(z) U(z)^{-a}$$

for all $z \in S^{2n-1}$, where $U(z) \in PU(d) = \mathrm{Aut}(M_d(\mathbb{C}))$ is the projective unitary given above, and e_+^{2n} (resp. e_-^{2n}) denotes the $2n$-dimensional northern (resp. southern) hemisphere.

Each d-homogeneous C^*-algebra over $S^{2n-1} \times S^1$ corresponding to each element of $[S^{2n-1} \times S^1, BPU(d)]$ can be constructed.

Lemma 1.1. *Every d-homogeneous C^*-algebra over $S^{2n-1} \times S^1$, whose homogeneous C^*-subalgebra restricted to the subspace S^{2n-1} of $S^{2n-1} \times S^1$ has the trivial bundle structure, is isomorphic to one of the following C^*-subalgebras $A_{d,a}$, $a \in \mathbb{Z}$ or $\mathbb{Z}_d$, of $C(S^{2n-1} \times [0,1], M_d(\mathbb{C}))$ given as follows: if $f \in A_{d,a}$, then the following condition is satisfied*

$$f(z,1) = U(z)^a f(z,0) U(z)^{-a}$$

for all $z \in S^{2n-1}$, where $U(z) \in PU(d)$ is the unitary given above.

Proof. Let A be a d-homogeneous C^*-algebra over $S^{2n-1} \times S^1$ whose homogeneous C^*-subalgebra restricted to the subspace S^{2n-1} of $S^{2n-1} \times S^1$ has the trivial bundle structure. Since there is a map of degree 1 from $S^{2n-1} \times S^1$ to S^{2n}, the composite of the map of degree 1 and the map representing each element of $[S^{2n}, BPU(d)]$ gives an element of $[S^{2n-1} \times S^1, BPU(d)]$. Hence each element of $[S^{2n}, BPU(d)] \cong [S^{2n-1}, PU(d)]$ representing a d-homogeneous C^*-algebra over S^{2n} induces an element of $[S^{2n-1}, PU(d)] \subset [S^{2n-1} \times S^1, BPU(d)]$, and the d-homogeneous C^*-algebras $A_{d,a}$ over $S^{2n-1} \times S^1$ corresponding to the d-homogeneous C^*-algebras $B_{d,a}$ over S^{2n} are constructed in the statement. By the hypothesis, the restriction of the d-homogeneous C^*-algebra A over $S^{2n-1} \times S^1$ to the subspace $S^{2n-1} \times (0,1)$ of $S^{2n-1} \times S^1$ has the trivial bundle structure. Hence A corresponds to an element of $[S^{2n-1}, PU(d)]$, and so A is characterized by the unitary $U(z)^a \in PU(d)$ over S^{2n-1} for some $a \in \mathbb{Z}$ or $\mathbb{Z}_d$.

For a d-homogeneous C^*-algebra A over S^{2n-1} there is a matrix algebra $M_q(\mathbb{C})$ such that $A \otimes M_q(\mathbb{C})$ is isomorphic to $C(S^{2n-1}) \otimes M_{dq}(\mathbb{C})$. Since there is a map of degree 1 from S^{2n+1} to $S^{2n} \times S^1$, there are d-homogeneous C^*-algebras over $S^{2n} \times S^1$ induced from d-homogeneous C^*-algebras over S^{2n+1}. And also there are d-homogeneous C^*-algebras over $S^{2n} \times S^1$ induced from d-homogeneous C^*-algebras over S^{2n}. But the tensor product of each d-homogeneous C^*-algebra over $S^{2n} \times S^1$ induced from a d-homogeneous C^*-algebra over S^{2n+1} with $M_q(\mathbb{C})$ has the trivial bundle structure for some integer q big enough since $[S^{2n+1}, BPU(dq)] \cong \{0\}$. And there is a map of degree 1 from S^{2n} to $S^{2n-1} \times S^1$, and so there are d-homogeneous C^*-algebras over $S^{2n-1} \times S^1$ induced from d-homogeneous C^*-algebras over S^{2n}. See [13, 15] for details.

From now on, we assume that every d-homogeneous C^*-algebra over $S^{2n} \times S^1$ is isomorphic to the tensor product of a d-homogeneous C^*-algebra over S^{2n} with $C(S^1)$.

Lemma 1.2. *Let n and k be integers greater than 1. Each d-homogeneous C^*-algebra over $S^n \times S^k$ is isomorphic to a d-homogeneous C^*-algebra characterized by the unitary $U(z)^a$ over S^{n-1} in a d-homogeneous C^*-algebra P_c over*

$e^n_+ \times S^k$ and $e^n_- \times S^k$, where $U(z) \in PU(d)$ or $PU(c)$ if $M_c(\mathbb{C})$ is factored out of P_c.

Proof. Since e^n_+ and e^n_- are contractible, each d-homogeneous C^*-algebra over $e^n_+ \times S^k$ and $e^n_- \times S^k$ is essentially induced by a d-homogeneous C^*-algebra over S^k. Each d-homogeneous C^*-algebra over $S^n \times S^k$ is characterized by a projective unitary over the boundaries $S^{n-1} \times S^k$ of $e^n_+ \times S^k$ and $e^n_- \times S^k$. But $\pi_1(S^n) = \{0\}$ and so the identification of the boundaries $S^k \hookrightarrow e^n_+ \times S^k$ and $S^k \hookrightarrow e^n_- \times S^k$ does give the trivial bundle structure. Hence the d-homogeneous C^*-algebra over $S^n \times S^k$ is characterized by the unitary $U(z)^a$, $a \in \mathbb{Z}$ or $a \in \mathbb{Z}_d$, over S^{n-1} in the d-homogeneous C^*-algebra over $e^n_+ \times S^k$ and $e^n_- \times S^k$, where $U(z) \in PU(d)$ or $PU(c)$.

Note that $[S^2 \times S^2, BPU(d)] \hookrightarrow H^2(S^2 \times S^2, \mathbb{Z}_d) \oplus H^4(S^2 \times S^2, \mathbb{Z})$ by the Woodward theorem [18], and that each d-homogeneous C^*-algebra over $S^2 \times S^2$ can be constructed by the trick of Lemma 1.2. Each d-homogeneous C^*-algebra over $S^2 \times S^2$ is induced from a d-homogeneous C^*-algebra over $S^2 \times \mathbb{T}^2$, since there is a map of degree 1 from $S^2 \times S^2$ to $S^2 \times \mathbb{T}^2$, and so d-homogeneous C^*-algebras over $S^2 \times S^2$ are essentially the same as d-homogeneous C^*-algebras over $S^2 \times \mathbb{T}^2$.

More generally, each d-homogeneous C^*-algebra over $\prod_{i=1}^e S^{2n_i} \times \mathbb{T}^2$ corresponding to each element of

$$\left[\prod_{i=1}^e S^{2n_i} \times \mathbb{T}^2, BPU(d) \right] \hookrightarrow H^{\text{even} \geq 2}\left(\prod_{i=1}^e S^{2n_i} \times \mathbb{T}^2, \mathbb{Z} \right)$$

can be constructed, where $H^{\text{even} \geq 2}$ denotes the direct sum of the even dimensional Čech cohomology groups, which are isomorphic to the even dimensional singular cohomology groups, since the Čech cohomology groups $H^n(M, \mathbb{Z})$ are isomorphic to the singular cohomology groups $H^n(M, \mathbb{Z})$ when M is *triangularizable* (see [5, Theorem 15.8]).

Theorem 1.3. *Let A_d be a d-homogeneous C^*-algebra over $\prod_{i=1}^e S^{2n_i} \times \mathbb{T}^2$, whose d-homogeneous C^*-subalgebra restricted to the subspace $\mathbb{T}^2$ of $\prod_{i=1}^e S^{2n_i} \times \mathbb{T}^2$ is isomorphic to a rational rotation algebra $A_{\frac{1}{d}}$. Then A_d is isomorphic to one of the following C^*-subalgebras $A^{a_1, a_2, \cdots, a_e}$, $a_1, \cdots, a_e \in \mathbb{Z}$, of*

$$C\left(\prod_{i=1}^e (e^{2n_i}_+ \amalg e^{2n_i}_-) \times \mathbb{T}^1 \times [0,1], M_d(\mathbb{C}) \right)$$

consisting of those functions f that satisfy

$$(f|_{e^{2n_i}_+ \amalg e^{2n_i}_-})+(z_i) = U(z_i)^{a_i}(f|_{e^{2n_i}_+ \amalg e^{2n_i}_-})-(z_i)U(z_i)^{-a_i}$$

$$(f|_{\mathbb{T}^1 \times [0,1]})(w, 1) = U(w)^l(f|_{\mathbb{T}^1 \times [0,1]})(w, 0)U(w)^{-l}$$

for all $(z_1, \cdots, z_e, w) \in \prod_{i=1}^{e} S^{2n_i - 1} \times \mathbb{T}^1$, where $U(z_i), U(w) \in PU(d)$ are as defined before.

Proof. By Lemma 1.2, replacing S^{2n_i} by S^2 does not give any change in the relationship, associated with bundle structure, among the factors of $\prod_{i=1}^{e} S^{2n_i} \times \mathbb{T}^2$. Hence each d-homogeneous C^*-algebra over $\prod_{i=1}^{e} S^{2n_i} \times \mathbb{T}^2$ can be given by [4, Theorem 2.5], which is exactly the same as the case $n_i = 1$ for all i.

Thomsen [17, Theorem 1.15] computed

$$\pi_{2c-1}(\mathrm{Aut}(M_{dp}(\mathbb{C}) \otimes M_{q^\infty})) \cong \mathbb{Z}/dp\mathbb{Z}$$

for M_{q^∞} a UHF-algebra of type q^∞, and dp and q relatively prime integers. The result clarifies the bundle structure of the tensor product of a d-homogeneous C^*-algebra over S^{2c} with M_{q^∞}. Let $A_{d,a}$ be a d-homogeneous C^*-algebra over $S^{2n-1} \times S^1$ of which no non-trivial matrix algebra can be factored out. This result implies that for any positive integer p any matrix algebra bigger than $M_p(\mathbb{C})$ is not factored out of $A_{d,a} \otimes M_p(\mathbb{C})$, $(d, a) = 1$, since if it were, then the same would be true for the d-homogeneous C^*-algebra $B_{d,a}$ over S^{2c} corresponding to $A_{d,a}$. But any matrix algebra bigger than $M_p(\mathbb{C})$ is not factored out of $B_{d,a} \otimes M_p(\mathbb{C})$, since if it were, then $\pi_{2c-1}(\mathrm{Aut}(M_{dp}(\mathbb{C}) \otimes M_{q^\infty})) \cong F \subsetneq \mathbb{Z}/dp\mathbb{Z}$. This contradicts the Thomsen result. So the natural inclusion $C(S^1) \hookrightarrow A_{d,a}$ induces the canonical homomorphism $K_0(C(S^1)) \to K_0(A_{d,a})$ such that $[1_{C(S^1)}]$ maps to $[1_{A_{d,a}}]$.

Lemma 1.4. *Let $A_{d,a}$ be a d-homogeneous C^*-algebra over $S^{2n-1} \times S^1$ of which no non-trivial matrix algebra can be factored out. Then $[1_{A_{d,a}}] \in K_0(A_{d,a}) \cong \mathbb{Z}^2$ is primitive.*

Proof. For M a compact CW-complex the Čech cohomology group $H^3(M, \mathbb{Z})$ classifies the tensor products of d-homogeneous C^*-algebras over M with the C^*-algebra $\mathcal{K}(\mathcal{H})$ of compact operators on a separable Hilbert space $\mathcal{H}$ (see [7]). $H^3(S^{2n-1} \times S^1, \mathbb{Z}) \cong \{0\}$ if $n \neq 2$, and $H^3(S^3 \times S^1, \mathbb{Z}) \cong H^3(S^3, \mathbb{Z}) \cong \mathbb{Z}$. But by the Woodward theorem[18] $[S^3 \times S^1, BPU(d)]$ is embedded into $H^2(S^3 \times S^1, \mathbb{Z}_d) \oplus H^4(S^3 \times S^1, \mathbb{Z}) \cong H^4(S^3 \times S^1, \mathbb{Z}) \cong \mathbb{Z}$. By Lemma 1.1 each d-homogeneous C^*-algebra $A_{d,a}$ over $S^3 \times S^1$ is induced from a d-homogeneous C^*-algebra $B_{d,a}$ over S^4. Since $H^3(S^4, \mathbb{Z}) \cong \{0\}$, $B_{d,a}$ is stably isomorphic to $C(S^4) \otimes M_d(\mathbb{C})$. So $A_{d,a}$ is stably isomorphic to $C(S^3 \times S^1) \otimes M_d(\mathbb{C})$. Hence $A_{d,a}$ is stably isomorphic to $C(S^{2n-1} \times S^1) \otimes M_d(\mathbb{C})$. Since $K_0(C(S^{2n-1} \times S^1)) \cong \mathbb{Z}^2$, $K_0(A_{d,a}) \cong \mathbb{Z}^2$. Hence it is enough to show that $[1_{A_{d,a}}] \in K_0(A_{d,a})$ is primitive.

No matrix algebra bigger than $M_q(\mathbb{C})$ can be factored out of $A_{d,a} \otimes M_q(\mathbb{C})$, and so $C(S^{2n-1})$ cannot be factored out of $A_{d,a} \otimes M_q(\mathbb{C})$. Hence the canonical imbedding ϕ of $C(S^{2n-1})$ into $A_{d,a}$ induces an isomorphism μ of $K_0(C(S^{2n-1} \times S^1))$ into $K_0(A_{d,a})$. But the unit $1_{C(S^{2n-1})}$ maps to the unit $1_{C(S^{2n-1} \times S^1)}$ under the canonical imbedding ψ of $C(S^{2n-1})$ into $C(S^{2n-1} \times S^1)$. Thus $[1_{C(S^{2n-1})}] \in K_0(C(S^{2n-1})) \cong \mathbb{Z}$ maps to $[1_{C(S^{2n-1} \times S^1)}] \in K_0(C(S^{2n-1} \times S^1)) \cong \mathbb{Z}^2$, primitive in $K_0(C(S^{2n-1} \times S^1))$ (see [12]). In the commutative diagram

$$K_0(C(S^{2n-1}))@>\psi_*>> K_0(C(S^{2n-1} \times S^1))$$

$$@V(\text{identity})_*VV @VV\mu(\cong)V$$

$$K_0(C(S^{2n-1}))@>\phi_*>> K_0(A_{d,a}),$$

$\mu([1_{C(S^{2n-1} \times S^1)}]) = \phi_* \circ (\text{identity})_* \circ \psi_*^{-1}([1_{C(S^{2n-1} \times S^1)}]) = [1_{A_{d,a}}]$, i.e., μ must be the canonical extension of $(\text{identity})_* : K_0(C(S^{2n-1})) \to K_0(C(S^{2n-1}))$. So $[1_{A_{d,a}}]$ is the image of the primitive element $[1_{C(S^{2n-1} \times S^1)}] \in K_0(C(S^{2n-1} \times S^1))$ under the isomorphism μ. Hence $[1_{A_{d,a}}] \in K_0(A_{d,a}) \cong \mathbb{Z}^2$ is primitive. Therefore, $K_0(A_{d,a}) \cong \mathbb{Z}^2$ and $[1_{A_{d,a}}] \in K_0(A_{d,a})$ is primitive.

Lemma 1.5. *Let $B_{d,a}$ be d-homogeneous C^*-algebra over S^{2n} of which no non-trivial matrix algebra can be factored out. Then $[1_{B_{d,a}}] \in K_0(B_{d,a}) \cong \mathbb{Z}^2$ is primitive.*

Proof. Since $H^3(S^{2n}, \mathbb{Z}) = \{0\}$ for any positive integer n, $B_{d,a}$ is stably isomorphic to $C(S^{2n}) \otimes M_d(\mathbb{C})$. So $K_0(B_{d,a}) \cong K_0(C(S^{2n})) \cong \mathbb{Z} \oplus \mathbb{Z}$. But $B_{d,a}$ corresponds to $A_{d,a}$ with respect to the condition on sections over the boundaries S^{2n-1} of $e_+^{2n} \amalg e_-^{2n}$ and $S^{2n-1} \times [0,1]$, and the canonical embedding of $C(S^{2n-1})$ into $A_{d,a}$ which induces the isomorphism of $K_0(C(S^{2n-1} \times S^1))$ into $K_0(A_{d,a})$ corresponds to the embedding ϕ of $C(S^{2n-1})$ into $B_{d,a}$, where $S^{2n-1} = \partial e_\pm^{2n}$. The canonical embedding ϕ of $C(S^{2n-1})$ into $B_{d,a}$ induces an isomorphism μ of $K_0(C(S^{2n})) \cong \mathbb{Z} \oplus \mathbb{Z}$ into $K_0(B_{d,a}) \cong \mathbb{Z} \oplus \mathbb{Z}$. The unit $1_{C(S^{2n-1})}$ maps to the unit $1_{C(S^{2n})}$ under the canonical embedding ψ of $C(S^{2n-1})$ into $C(S^{2n})$. $[1_{C(S^{2n-1})}] \in K_0(C(S^{2n-1})) \cong \mathbb{Z}$ maps to $[1_{C(S^{2n})}] \in K_0(C(S^{2n})) \cong \mathbb{Z}^2$, primitive in $K_0(C(S^{2n}))$ (see [12]). In the commutative diagram

$$K_0(C(S^{2n-1}))@>\psi_*>> K_0(C(S^{2n}))$$

$$@V(\text{identity})_*VV @VV\mu(\cong)V$$

$$K_0(C(S^{2n-1}))@>\phi_*>> K_0(B_{d,a}),$$

$\mu([1_{C(S^{2n})}]) = \phi_* \circ (\text{identity})_* \circ \psi_*^{-1}([1_{C(S^{2n})}]) = [1_{B_{d,a}}]$. That is, μ must be the canonical extension of $(\text{identity})_* : K_0(C(S^{2n-1})) \to K_0(C(S^{2n-1}))$. So $[1_{B_{d,a}}]$ is the image of the primitive element $[1_{C(S^{2n})}] \in K_0(C(S^{2n}))$ under the isomorphism μ. Hence $[1_{B_{d,a}}] \in K_0(B_{d,a})$ is primitive. Therefore, $[1_{B_{d,a}}] \in K_0(B_{d,a}) \cong \mathbb{Z}^2$ is primitive.

Theorem 1.6. *Let A_d be a C^*-algebra over $\prod_{i=1}^{e} S^{2n_i} \times \mathbb{T}^2$ defined in Theorem 1.3. Then $K_0(A_d) \cong K_1(A_d) \cong \mathbb{Z}^{2^{e+1}}$, and $[1_{A_d}] \in K_0(A_d)$ is primitive.*

Proof. We will show later that A_d is stably isomorphic to $C(\prod_{i=1}^{e} S^{2n_i} \times \mathbb{T}^2) \otimes M_d(\mathbb{C})$. By the Künneth theorem

$$
K_0(A_d) \cong K_0\left(C\left(\prod_{i=1}^{e} S^{2n_i} \times \mathbb{T}^2 \right) \right)
$$

$$
\cong K_0\left(C\left(\prod_{i=1}^{e} S^{2n_i} \right) \right) \otimes K_0(C(\mathbb{T}^2)) \oplus K_1\left(C\left(\prod_{i=1}^{e} S^{2n_i} \right) \right)
$$

$$
\otimes K_1(C(\mathbb{T}^2))
$$

$$
\cong \mathbb{Z}^{2^e} \otimes \mathbb{Z}^2 \oplus \{0\} \cong \mathbb{Z}^{2^{e+1}}.
$$

Similarly, one obtains that $K_1(A_d) \cong \mathbb{Z}^{2^{e+1}}$.

It is enough to show that $[1_{A_d}] \in K_0(A_d)$ is primitive. By the same reasoning as the proof given in Lemma 1.5, the canonical embedding ϕ of $C(\prod_{i=1}^{e} S^{2n_i-1} \times \mathbb{T}^1)$ into A_d induces an isomorphism μ of $K_0(C(\prod_{i=1}^{e} S^{2n_i} \times \mathbb{T}^2))$ into $K_0(A_d)$. The unit $1_{C(\prod_{i=1}^{e} S^{2n_i-1} \times \mathbb{T}^1)}$ maps to the unit $1_{C(\prod_{i=1}^{e} S^{2n_i} \times \mathbb{T}^2)}$ under the canonical embedding ψ of $C(\prod_{i=1}^{e} S^{2n_i-1} \times \mathbb{T}^1)$ into $C(\prod_{i=1}^{e} S^{2n_i} \times \mathbb{T}^2)$. So

$$
\left[1_{C(\prod_{i=1}^{e} S^{2n_i-1} \times \mathbb{T}^1)} \right] \in K_0\left(C\left(\prod_{i=1}^{e} S^{2n_i-1} \times \mathbb{T}^1 \right) \right)
$$

maps to

$$
\left[1_{C(\prod_{i=1}^{e} S^{2n_i} \times \mathbb{T}^2)} \right] \in K_0\left(C\left(\prod_{i=1}^{e} S^{2n_i} \times \mathbb{T}^2 \right) \right),
$$

primitive in $K_0(C(\prod_{i=1}^{e} S^{2n_i} \times \mathbb{T}^2))$. In the commutative diagram

$$
K_0(C(\prod_{i=1}^{e} S^{2n_i-1} \times \mathbb{T}^1)) @>\psi_*>> K_0(C(\prod_{i=1}^{e} S^{2n_i} \times \mathbb{T}^2))
$$

$$
@V(\text{identity})_* VV @VV\mu(\cong)V
$$

$$
K_0(C(\prod_{i=1}^{e} S^{2n_i-1} \times \mathbb{T}^1)) @>\phi_*>> K_0(A_d).
$$

$$
\mu\left(\left[1_{C(\prod_{i=1}^{e} S^{2n_i} \times \mathbb{T}^2)} \right] \right) = \phi_* \circ (\text{identity})_* \circ \psi_*^{-1}\left(\left[1_{C(\prod_{i=1}^{e} S^{2n_i} \times \mathbb{T}^2)} \right] \right) = [1_{A_d}],
$$

i.e., μ must be the canonical extension of $(\text{identity})_*$:

$$
K_0\left(C\left(\prod_{i=1}^{e} S^{2n_i-1} \times \mathbb{T}^1 \right) \right) \to K_0\left(C\left(\prod_{i=1}^{e} S^{2n_i-1} \times \mathbb{T}^1 \right) \right).
$$

So $[1_{A_d}]$ is the image of the primitive element

$$\left[1_{C(\prod_{i=1}^{e} S^{2n_i} \times \mathbb{T}^2)}\right] \in K_0\left(C\left(\prod_{i=1}^{e} S^{2n_i} \times \mathbb{T}^2\right)\right)$$

under the isomorphism μ. Hence $[1_{A_d}] \in K_0(A_d)$ is primitive.

Therefore, $K_0(A_d) \cong K_1(A_d) \cong \mathbb{Z}^{2^{e+1}}$, and $[1_{A_d}] \in K_0(A_d)$ is primitive.

2. C^*-algebras over Spheres with Fibres Noncommutative Tori of Rank 3

Definition 2.1. Let A_d be a d-homogeneous C^*-algebra over $\prod_{i=1}^{e} S^{2n_i} \times \mathbb{T}^2$ whose d-homogeneous C^*-subalgebra restricted to the subspace $\mathbb{T}^2$ of $\prod_{i=1}^{e} S^{2n_i} \times \mathbb{T}^2$ is a rational rotation algebra $A_{\frac{1}{d}}$. The *spherical noncommutative torus* $\mathbb{S}_\rho^d$ *of rank* $e+1$ is given by a crossed product by an action α of $\mathbb{Z}$ on A_d, where the action α of $\mathbb{Z}$ on $C(\prod_{i=1}^{e} S^{2n_i}) \otimes M_d(\mathbb{C})$ is trivial, and $C(\mathbb{T}^2) \times_\alpha \mathbb{Z}$ is a completely irrational noncommutative torus $A_\rho \cong C^*(\widehat{\mathbb{T}^2} \times \mathbb{Z}, \rho)$ of rank 3 for a totally skew multiplier ρ on $\widehat{\mathbb{T}^2} \times \mathbb{Z}$.

Then the fibre of $\mathbb{S}_\rho^d$, which is called a *generalized noncommutative torus of rank* 3 and denoted by P_ρ^d, is isomorphic to the crossed product by the action α of $\mathbb{Z}$ on the rational rotation algebra $A_{\frac{1}{d}}$, where the action α on the fibre $M_d(\mathbb{C})$ of $A_{\frac{1}{d}}$ is trivial. Thus the spherical noncommutative torus $\mathbb{S}_\rho^d$ is realized as the C^*-algebra of sections of a locally trivial C^*-algebra bundle over $\prod_{i=1}^{e} S^{2n_i}$ with fibres P_ρ^d.

Elliott [8, Theorem 2.2] proved that, for a noncommutative torus A_ρ of rank m, $K_0(A_\rho) \cong K_1(A_\rho) \cong \mathbb{Z}^{2^{m-1}}$ and $[1_{A_\rho}] \in K_0(A_\rho)$ is primitive.

We are going to show that $[1_{\mathbb{S}_\rho^d}] \in K_0(\mathbb{S}_\rho^d)$ is primitive.

Theorem 2.2. *Let $\mathbb{S}_\rho^d$ be a spherical noncommutative torus of rank $e+1$ defined above. Then $K_0(\mathbb{S}_\rho^d) \cong K_1(\mathbb{S}_\rho^d) \cong \mathbb{Z}^{2^{e+2}}$ and $[1_{\mathbb{S}_\rho^d}] \in K_0(\mathbb{S}_\rho^d)$ is primitive.*

Proof. It was shown in Theorem 1.6 that $K_0(A_d) \cong K_1(A_d) \cong \mathbb{Z}^{2^{e+1}}$ and $[1_{A_d}] \in K_0(A_d)$ is primitive. Write $\mathbb{S}_\rho^d = C^*(A_d, u_3)$. Also, we can think of $\mathbb{S}_\rho^d$ as the crossed product of A_d by an action α of $\mathbb{Z}$, where the generator of $\mathbb{Z}$ corresponds to u_3, which acts on $C^*(\widehat{\mathbb{T}^2}) \cong C^*(u_1^d, u_2^d)$ by conjugation (sending u_j^d to $u_3 u_j^d u_3^{-1} = e^{2\pi i d\theta_{j3}} u_j^d, j = 1, 2$), and which acts trivially on $C(\prod_{i=1}^{e} S^{2n_i}) \otimes M_d(\mathbb{C})$. Notice that this action is homotopic to the trivial action, since we can homotope θ_{j3} to 0. Hence $\mathbb{Z}$ acts trivially on the K-theory of A_d. The Pimsner-Voiculescu exact sequence for a crossed product gives an exact sequence

$$K_0(A_d)@ > 1 - \alpha_* >> K_0(A_d)@ > \Phi >> K_0(\mathbb{S}_\rho^d)@$$
$$>>> K_1(A_d)@ > 1 - \alpha_* >> K_1(A_d)$$

and, similarly, for K_1, where the map Φ is induced by inclusion. Since $\alpha_* = 1$ and since the K-groups of A_d are free abelian, this reduces a split short exact sequence

$$\{0\}@ >>> K_0(A_d)@ > \Phi >> K_0(\mathbb{S}_\rho^d)@ >>> K_1(A_d)@ >>> \{0\}$$

and, similarly, for K_1. So $K_0(\mathbb{S}_\rho^d)$ and $K_1(\mathbb{S}_\rho^d)$ are free abelian groups of rank $2 \cdot 2^{e+1} = 2^{e+2}$. Furthermore, since the inclusion $A_d \to \mathbb{S}_\rho^d$ sends 1_{A_d} to $1_{\mathbb{S}_\rho^d}$, $[1_{\mathbb{S}_\rho^d}]$ is the image of $[1_{A_d}]$, which is primitive in $K_0(A_d)$. Hence the image is primitive, since the Pimsner-Voiculescu exact sequence is a split short exact sequence of torsion-free groups. Therefore, $K_0(\mathbb{S}_\rho^d) \cong K_1(\mathbb{S}_\rho^d) \cong \mathbb{Z}^{2^{e+2}}$, and $[1_{\mathbb{S}_\rho^d}] \in K_0(\mathbb{S}_\rho^d)$ is primitive.

Lemma 2.3. *Each d-homogeneous C^*-algebra over $\prod_{i=1}^e S^{2n_i} \times \mathbb{T}^2$ is stably isomorphic to $C(\prod_{i=1}^e S^{2n_i} \times \mathbb{T}^2) \otimes M_d(\mathbb{C})$.*

Proof. Each non-trivial element in the Čech (=singular) cohomology group $H^3(\prod_{i=1}^e S^{2n_i} \times \mathbb{T}^2, \mathbb{Z})$ can be given by a non-trivial element in $H^3(S^2 \times \mathbb{T}^1, \mathbb{Z}) \cong \mathbb{Z}$ if there exist such factors. But by the Woodward theorem [18], $[S^2 \times \mathbb{T}^1, BPU(d)]$ is embedded into $H^2(S^2 \times \mathbb{T}^1, \mathbb{Z}_d) \oplus H^4(S^2 \times \mathbb{T}^1, \mathbb{Z}) \cong H^2(S^2, \mathbb{Z}_d) \cong \mathbb{Z}_d$. So each d-homogeneous C^*-algebra over $S^2 \times \mathbb{T}^1$ is isomorphic to the tensor product of a d-homogeneous C^*-algebra over S^2 with $C(\mathbb{T}^1)$, which is stably isomorphic to $C(S^2) \otimes C(\mathbb{T}^1) \otimes M_d(\mathbb{C})$, since $H^3(S^2, \mathbb{Z}) = \{0\}$. Thus each d-homogeneous C^*-algebra over $S^2 \times \mathbb{T}^1$ is stably isomorphic to $C(S^2 \times \mathbb{T}^1) \otimes M_d(\mathbb{C})$. Therefore, every d-homogeneous C^*-algebra over $\prod_{i=1}^e S^{2n_i} \times \mathbb{T}^2$ is stably isomorphic to $C(\prod_{i=1}^e S^{2n_i} \times \mathbb{T}^2) \otimes M_d(\mathbb{C})$.

We are going to show that $\mathbb{S}_\rho^d \otimes \mathcal{K}(\mathcal{H})$ has the trivial bundle structure.

Theorem 2.4. *The spherical noncommutative torus $\mathbb{S}_\rho^d$, as defined before, is stably isomorphic to $C(\prod_{i=1}^e S^{2n_i}) \otimes A_\rho \otimes M_d(\mathbb{C})$. In particular, P_ρ^d is stably isomorphic to $A_\rho \otimes M_d(\mathbb{C})$.*

Proof. By Lemma 2.3, the d-homogeneous C^*-algebra A_d is stably isomorphic to $C(\prod_{i=1}^e S^{2n_i} \times \mathbb{T}^2) \otimes M_d(\mathbb{C})$. In particular, $C(\prod_{i=1}^e S^{2n_i})$ is factored out of $A_d \otimes \mathcal{K}(\mathcal{H})$. By the definition of $\mathbb{S}_\rho^d$, $C(\prod_{i=1}^e S^{2n_i})$ is factored out of $\mathbb{S}_\rho^d \otimes \mathcal{K}(\mathcal{H})$.

So $\mathbb{S}_\rho^d$ is stably isomorphic to $C(\prod_{i=1}^e S^{2n_i}) \otimes P_\rho^d$. But P_ρ^d is realized as the crossed product $A_{\frac{l}{d}} \times_\alpha \mathbb{Z}$, where α acts trivially on the fibre $M_d(\mathbb{C})$ of $A_{\frac{l}{d}}$. So

$$P_\rho^d \otimes \mathcal{K}(\mathcal{H}) \cong (A_{\frac{l}{d}} \times_\alpha \mathbb{Z}) \otimes \mathcal{K}(\mathcal{H}) \cong (A_{\frac{l}{d}} \otimes \mathcal{K}(\mathcal{H})) \times_{\tilde{\alpha}} \mathbb{Z},$$

where $\tilde{\alpha}$ is the canonical extension of α such that $\tilde{\alpha}$ acts trivially on $M_d(\mathbb{C}) \otimes \mathcal{K}(\mathcal{H})$. So

$$P_\rho^d \otimes \mathcal{K}(\mathcal{H}) \cong (C(\mathbb{T}^2) \otimes M_d(\mathbb{C}) \otimes \mathcal{K}(\mathcal{H})) \times_{\tilde{\alpha}} \mathbb{Z}$$
$$\cong (C(\mathbb{T}^2) \times_\alpha \mathbb{Z}) \otimes M_d(\mathbb{C}) \otimes \mathcal{K}(\mathcal{H}).$$

Thus P_ρ^d is stably isomorphic to $A_\rho \otimes M_d(\mathbb{C})$. Therefore, $\mathbb{S}_\rho^d$ is stably isomorphic to $C(\prod_{i=1}^e S^{2n_i}) \otimes A_\rho \otimes M_d(\mathbb{C})$.

The completely irrational noncommutative torus A_ρ in Definition 2.1 is realized as $C^*(u_1, u_2, u_3 \mid u_i u_2 = u_2 u_1, u_1 u_3 = e^{2\pi i \theta} u_3 u_1, u_2 u_3 = e^{2\pi i \beta} u_3 u_2)$. Then $\mathrm{tr}(K_0(A_\rho)) = \mathbb{Z} + \mathbb{Z} + \theta\mathbb{Z} + \beta\mathbb{Z}$ and $\mathrm{tr}(K_0(P_\rho^d)) = \frac{1}{d}(\mathbb{Z} + \mathbb{Z} + \theta\mathbb{Z} + \beta\mathbb{Z})$ (see [15, Lemma 4.1]). It was shown in [2, Theorem 1.5] that every completely irrational noncommutative torus A_ω has real rank 0, where the "real rank 0" means that the set of invertible self-adjoint elements is dense in the set of self-adjoint elements. Combining Theorem 2.4 and [6, Corollary 3.3] yields that P_ρ^d has real rank 0. The Lin and Rørdam theorem [14, Proposition 3] says that P_ρ^d is an inductive limit of circle algebras, since $P_\rho^d \otimes \mathcal{K}(\mathcal{H}) \cong A_\rho \otimes \mathcal{K}(\mathcal{H})$ is an inductive limit of circle algebras [14, Proposition 2]. Let A_ω be realized as $C^*(u, v, w \mid uv = e^{2\pi i \frac{l}{d}} vu, uw = e^{2\pi i \frac{\theta}{d}} wu, vw = e^{2\pi i \frac{\beta}{d}} wv)$. Combining [9, Theorem 7.1] and [11, Theorem 1.3] yields that the completely irrational noncommutative torus A_ω of rank 3 and the generalized noncommutative torus P_ρ^d of rank 3 are isomorphic, since $\mathrm{tr}(K_0(A_\omega)) = \mathbb{Z} + \frac{l}{d}\mathbb{Z} + \frac{\theta}{d}\mathbb{Z} + \frac{\beta}{d}\mathbb{Z} = \mathrm{tr}(K_0(P_\rho^d))$.

It was shown in [15, Proposition 5.2] that if a completely irrational noncommutative torus A_ω of rank 3 cannot be realized as a crossed product by an action α of $\mathbb{Z}$ on a non-trivial rational rotation algebra $A_{\frac{l}{d}}$ then each C^*-algebra of sections of a locally trivial C^*-algebra bundle over $\prod_{i=1}^e S^{2n_i}$ with fibres A_ω has the trivial bundle structure. So each C^*-algebra of sections of a locally trivial C^*-algebra bundle over $\prod_{i=1}^e S^{2n_i}$ with fibres A_ρ has the trivial bundle structure. Hence each C^*-algebra of sections of a locally trivial C^*-algebra bundle over $\prod_{i=1}^e S^{2n_i}$ with fibres $P_\rho^d \cong A_\omega$ is isomorphic to a spherical noncommutative torus $\mathbb{S}_\rho^d$ of rank $e+1$ with primitive ideal space $\prod_{i=1}^e S^{2n_i}$.

Theorem 2.5. *All C^*-algebras of sections of locally trivial C^*-algebra bundles over $\prod_{i=1}^e S^{2n_i}$ with fibres completely irrational noncommutative tori A_ω of rank 3 are classified by $(\mathbb{Z}^e, \mathrm{tr}(K_0(A_\omega)))$ if $A_\omega \cong P_\rho^d$ for $d \neq 1$, and by $\mathrm{tr}(K_0(A_\omega))$ if there does not exist an integer $d > 1$ such that $A_\omega \cong P_\rho^d$.*

Proof. Assume that there does not exist an integer $d > 1$ such that $A_\omega \cong P_\rho^d$. Since each C^*-algebra of sections of a locally trivial C^*-algebra bundle over $\prod_{i=1}^e S^{2n_i}$ with fibres A_ω has the trivial bundle structure, combining [9, Theorem 7.1] and [11, Theorem 1.3] yields that all C^*-algebras of sections of locally trivial C^*-algebra bundles over $\prod_{i=1}^e S^{2n_i}$ with fibres completely irrational noncommutative tori A_ω of rank 3 are classified by $\mathrm{tr}(K_0(A_\omega))$.

Next, assume that $A_\omega \cong P_\rho^d$ for $d \neq 1$. Each C^*-algebra of sections of a locally trivial C^*-algebra bundle over $\prod_{i=1}^e S^{2n_i}$ with fibres $P_\rho^d \cong A_\omega$ is isomorphic to a spherical noncommutative torus $\mathbb{S}_\rho^d$ of rank $e + 1$ with primitive ideal space $\prod_{i=1}^e S^{2n_i}$. By Theorem 1.3, each d-homogeneous C^*-algebra A_d, whose d-homogeneous C^*-subalgebra restricted to the subspace $\mathbb{T}^2$ of $\prod_{i=1}^e S^{2n_i} \times \mathbb{T}^2$ is isomorphic to a rational rotation algebra $A_{\frac{1}{d}}$, is isomorphic to one of the C^*-algebras $A^{a_1, a_2, \cdots, a_e}$, $a_1, \cdots, a_e \in \mathbb{Z}$ given in the statement of Theorem 1.3. So they are classified by $\mathbb{Z}^e$. And combining [9, Theorem 7.1] and [11, Theorem 1.3] yields that the fibres P_ρ^d are classified by $\mathrm{tr}(K_0(P_\rho^d)) = \mathrm{tr}(K_0(A_\omega))$. Hence all C^*-algebras of sections of locally trivial C^*-algebra bundles over $\prod_{i=1}^e S^{2n_i}$ with fibres completely irrational noncommutative tori $A_\omega \cong P_\rho^d$ of rank 3 are classified by $(\mathbb{Z}^e, \mathrm{tr}(K_0(A_\omega)))$. Therefore, all C^*-algebras of sections of locally trivial C^*-algebra bundles over $\prod_{i=1}^e S^{2n_i}$ with fibres completely irrational noncommutative tori A_ω of rank 3 are classified by $(\mathbb{Z}^e, \mathrm{tr}(K_0(A_\omega)))$ if $A_\omega \cong P_\rho^d$ for $d \neq 1$, and by $\mathrm{tr}(K_0(A_\omega))$ if there does not exist an integer $d > 1$ such that $A_\omega \cong P_\rho^d$.

REFERENCES

1. L. Baggett and A. Kleppner, *Multiplier representations of abelian groups*, J. Funct. Anal. **14** (1973), 299–324.

2. B. Blackadar, A. Kumjian and M. Rørdam, *Approximately central matrix units and the structure of non-commutative tori*, K-Theory **6** (1992), 267–284.

3. F. Boca, *The structure of higher-dimensional noncommutative tori and metric Diophantine approximation*, J. Reine Angew. Math. **492** (1997), 179–219.

4. D. Boo, P. Kang and C. Park, *The sectional C^*-algebras over a torus with fibres a noncommutative torus*, Far East J. Math. Sci. **1** (1999), 561–579.

5. R. Bott and L.W. Tu, *Differential Forms in Algebraic Topology*, Springer-Verlag, New York, Heidelberg and Berlin, 1982.

6. L. Brown and G. Pedersen, *C^*-algebras of real rank zero*, J. Funct. Anal. **99** (1991), 131–149.

7. J. Dixmier, *C^*-Algebras*, North-Holland, Amsterdam, New York and Oxford, 1977.

8. G.A. Elliott, *On the K-theory of the C^*-algebra generated by a projective representation of a torsion-free discrete abelian group*, Operator Algebras and Group Representations (G. Aresene et al., ed.), vol. 1, Pitman, London, 1984, pp. 157–184.

9. ______, *On the classification of C^*-algebras of real rank zero*, J. Reine Angew. Math. **443** (1993), 179–219.

10. P. Green, *The local structure of twisted covariance algebras*, Acta Math. **140** (1978), 191–250.

11. R. Ji and J. Xia, *On the classification of commutator ideals*, J. Funct. Anal. **78** (1988), 208–232.

12. M. Karoubi, *K-Theory*, Springer-Verlag, Berlin, Heidelberg and New York, 1978.

13. F. Krauss and T.C. Lawson, *Examples of homogeneous C^*-algebras*, Memoirs A.M.S. **148** (1974), 153–164.

14. H. Lin and M. Rørdam, *Extensions of inductive limits of circle algebras*, J. London Math. Soc. **51** (1995), 603–613.

15. S. Oh and C. Park, *C^*-algebras over spheres with fibres noncommutative tori*, Proc. US-Japan Seminar on Operator Algebras and Applications (B. Blackadar and H. Kosaki, ed.) (to appear).

16. D. Poguntke, *The structure of twisted convolution C^*-algebras on abelian groups*, J. Operator Theory **38** (1997), 3–18.

17. K. Thomsen, *The homotopy type of the group of automorphisms of a UHF-algebra*, J. Funct. Anal. **72** (1987), 182–207.

18. L. Woodward, *The classification of principal $PU(k)$-bundles over a 4-complex*, J. London Math. Soc. **25** (1987), 513–524.

HOW BIG IS THE UNIFORM CONVERGENCE INTERVAL OF THE STRONG LAW OF LARGE NUMBERS?

QI-MAN SHAO[1]

ABSTRACT. Let $\{X_n, n \geq 1\}$ be i.i.d. r.v.'s with mean μ. A necessary and sufficient condition is given for the biggest interval in which the strong law of large numbers holds uniformly.

I. Introduction

Throughout this note let $\{X, X_n, n \geq 1\}$ be a sequence of independent and identically distributed nondegenerate random variables (i.i.d.r.v.'s) with mean μ. Put $S_0 = 0$, $S_n = \sum_{i=1}^{n} X_i$, $n = 1, 2, \cdots$. It is well-known by the Kolmogorov strong law of large numbers that except a null set

$$(1.1) \qquad \lim_{n \to \infty} (S_{n+k} - S_k)/n = \mu$$

for all $k \geq 0$. A natural question one would like to ask is if the convergence in (1.1) is uniformly in k. Unfortunately, the answer is negative, as shown by Sun (1993): (1.1) holds uniformly in $0 \leq k < \infty$ with probability zero. So, it would be interesting to know what the biggest uniform convergence interval of (1.1) could be. The main aim of the note is to give a necessary and sufficient condition for this.

Theorem 1.1. *Let $c > 0$ and $r \geq 1$. Then*

$$(1.2) \qquad P\left(\lim_{n \to \infty} (S_{n+k} - S_k)/n = \mu \ \ \textit{uniformly in } 0 \leq k \leq cn^r \right) = 1$$

if and only if $E|X|^r < \infty$.

Received November 9, 2000.

1991 *Mathematics Subject Classification.* 60F15.

Key Words and Phrases. uniform convergence interval, law of large numbers.

The research is partially supported by NSF grant DMS-9802451.

[1] Department of Mathematics, University of Oregon, Eugene, OR 97403, USA. Email: shao@math.uoregon.edu

Theorem 1.2. *Let $0 < \theta \le 1$. Then*

$$(1.3) \qquad P\left(\lim_{n\to\infty} (S_{n+k} - S_k)/n = \mu \ \ \text{uniformly in } 0 \le k \le e^{\varepsilon_n\, n^\theta} \right) = 1$$

for any positive sequence $\{\varepsilon_n, n \ge 1\}$ with $\varepsilon_n = o(1)$ if and only if $Ee^{t_0|X|^\theta} < \infty$ for some $t_0 > 0$.

The next theorem shows that the biggest possible uniform convergence interval cannot exceed $[0, e^{o(n)}]$ for any random variable.

Theorem 1.3. *For any $c > 0$,*

$$(1.4) \qquad P\left(\lim_{n\to\infty} (S_{n+k} - S_k)/n = \mu \ \ \text{uniformly in } 0 \le k \le e^{cn} \right) = 0.$$

The proof of Theorems 1.1~1.3 will be given in the next section, while generalizations of these results are discussed at the end of the paper.

2. Proofs

In what follows, without loss of generality, assume $\mu = 0$. The proofs are based on the following observation:

Lemma 2.1. *Let $a_x = a(x) \ge 0$ be a non-decreasing function. If*

$$(2.1) \qquad P\left(\lim_{n\to\infty} |S_{n+k} - S_k|/n = 0 \ \text{uniformly in } 0 \le k \le a_n \right) = 1, \quad \text{a.s.,}$$

then, for any $\varepsilon > 0$,

$$(2.2) \qquad\qquad E(|X| + a(|X|/\varepsilon)) < \infty.$$

Proof. If $\lim_{x\to\infty} a_x < \infty$, (2.2) follows from the law of large numbers. When $\lim_{x\to\infty} a_x = \infty$, without loss of generality, assume $a_1 \ge 2$. It is easy to see that

$$\left\{ \lim_{n\to\infty} |S_{n+k} - S_k|/n = 0 \ \ \text{uniformly in } 0 \le k \le a_n \right\}$$
$$= \left\{ \lim_{n\to\infty} \max_{0 \le k \le a_n} |S_{n+k} - S_k|/n = 0 \right\}$$

and

$$\max_{n-1+a_{n-1} < i \le n+a_n} |X_i| \le \max_{n+1 \le i \le n+a_n} |X_i|$$
$$\le \max_{0 \le k \le a_{n+1}} |S_{n+1+k} - S_k| + \max_{0 \le k \le a_n} |S_{n+k} - S_k|.$$

Thus, it follows from (2.1) that

$$\lim_{n\to\infty} \max_{n-1+a_{n-1}<i\leq n+a_n} |X_i|/n = 0 \quad \text{a.s.},$$

which yields immediately, by the well-known Borel-Cantelli lemma, for any $\varepsilon > 0$,

$$(2.3) \qquad \sum_{n=1}^{\infty} P\left(\max_{n-1+a_{n-1}<i\leq n+a_n} |X_i| > \varepsilon n \right) < \infty.$$

Clearly, (2.3) implies $P\left(\max_{n-1+a_{n-1}<i\leq n+a_n} |X_i| > \varepsilon n \right) \to 0$ as $n \to \infty$. Since $\{X, X_i, i \geq 1\}$ are i.i.d.r.v.'s, we have

$$P\left(\max_{n-1+a_{n-1}<i\leq n+a_n} |X_i| > \varepsilon n \right) \geq (1/2) \sum_{n-1+a_{n-1}<i\leq n+a_n} P(|X_i| > \varepsilon n)$$

$$\geq (1/4)\left(1 + a_n - a_{n-1}\right) P(|X| > \varepsilon n)$$

for every n sufficiently large. Hence, for any $\epsilon > 0$,

$$(2.4) \qquad \sum_{n=1}^{\infty} (1 + a_n - a_{n-1})\, P(|X| \geq \varepsilon (n-1)) < \infty,$$

which gives (2.2) via elementary arguments.

Proof of Theorem 1.1. The only if part follows from Lemma 2.1. We next prove that (1.2) is true if $E|X|^r < \infty$. Let

$$\widehat{X}_i = X_i I\{|X_i| \leq i^{1/r}\} - E X_i I\{|X_i| \leq i^{1/r}\}, \quad \widehat{S}_n = \sum_{i=1}^{n} \widehat{X}_i.$$

Note that

$$P\left(|X_n| > n^{1/r}, \ i.o.\right) = 0$$

and

$$\max_{0\leq k\leq cn^r} \sum_{i=k+1}^{k+n} |EX_i I\{|X_i| \leq i^{1/r}\}| \leq \sum_{i=1}^{n} E|X|I\{|X| > i^{1/r}\} = o(n).$$

It suffices to show that

$$(2.5) \qquad \limsup_{n\to\infty} \max_{0\leq k\leq cn^r} |\widehat{S}_{n+k} - \widehat{S}_k|/n = 0 \quad \text{a.s.}$$

Observe that

$$
\begin{aligned}
&\limsup_{n\to\infty} \max_{0\le k\le c\,n^r} |\widehat{S}_{n+k} - \widehat{S}_k|/n| \\
&\le \limsup_{m\to\infty} \max_{2^{m-1}\le n\le 2^m} \max_{0\le k\le c\,n^r} |\widehat{S}_{n+k} - \widehat{S}_k|/n \\
&\le 2\limsup_{m\to\infty} \max_{n\le 2^m} \max_{0\le k\le c\,2^{mr}} |\widehat{S}_{n+k} - \widehat{S}_k|/2^m \\
&\le 6\limsup_{m\to\infty} \max_{0\le i\le 1+c\,2^{m(r-1)}} \max_{n\le 2^m} |\widehat{S}_{i\,2^m+n} - \widehat{S}_{i\,2^m}|/2^m.
\end{aligned}
\tag{2.6}
$$

Let

$$
p = \begin{cases} 2 & \text{if } 1 \le r < 2, \\ 2r & \text{if } r \ge 2. \end{cases}
$$

By Rosenthal's inequality, we have, for any $\varepsilon > 0$

$$
\begin{aligned}
&P\left(\max_{0\le i\le 1+c\,2^{m(r-1)}} \max_{n\le 2^m} |\widehat{S}_{i\,2^m+n} - \widehat{S}_{i\,2^m}| \ge \varepsilon\,2^m \right) \\
&\le \sum_{0\le i\le 1+c\,2^{m(r-1)}} (\varepsilon\,2^m)^{-p} E \max_{n\le 2^m} |\widehat{S}_{i\,2^m+n} - \widehat{S}_{i\,2^m}|^p \\
&\le K(\varepsilon\,2^m)^{-p} \sum_{0\le i\le 1+c\,2^{m(r-1)}} \left\{ \left(\sum_{j=1+i2^m}^{(i+1)2^m} E|X_j|^2 I\{|X_j| \le j^{1/r}\} \right)^{p/2} \right. \\
&\qquad \left. + \sum_{j=1+i2^m}^{(i+1)2^m} E|X_j|^p I\{|X_j| \le j^{1/r}\} \right\} \\
&\le K(\varepsilon\,2^m)^{-p} 2^{m(r-1)} \left\{ 2^{mp/2} (E|X|^2 I\{|X| \le (1+c)2^m\})^{p/2} \right. \\
&\qquad \left. + 2^m E|X|^p I\{|X| \le (1+c)2^m\} \right\}.
\end{aligned}
$$

Here, and in the sequel, K denotes a universal constant, but its value is not necessarily the same at each occurrence. When $1 \le r < 2$, elementary arguments show

$$
\sum_{m=1}^{\infty} 2^{-m2}\, 2^{m(r-1)}\, 2^m\, E|X|^2 I\{|X| \le (1+c)2^m\} \le K(1 + E|X|^r) < \infty.
$$

Similarly, if $r \geq 2$, then we have

$$\sum_{m=1}^{\infty} 2^{-mp} 2^{m(r-1)} \left\{ 2^{mp/2} \left(E|X|^2 I\{|X| \leq (1+c)2^m\} \right)^{p/2} \right.$$
$$\left. + 2^m E|X|^p I\{|X| \leq (1+c)2^m\} \right\}$$
$$\leq \sum_{m=1}^{\infty} 2^{-mp} 2^{m(r-1)} 2^{mp/2} (E|X|^2)^{p/2} + \sum_{m=1}^{\infty} 2^{-mp} 2^{mr} E|X|^p I\{|X| \leq 2^m\}$$
$$< \infty.$$

Therefore, by the Borel-Cantelli lemma,

$$(2.7) \qquad \limsup_{m \to \infty} \max_{0 \leq i \leq 1+c\, 2^{m(r-1)}} \max_{n \leq 2^m} |\widehat{S}_{i\,2^m+n} - \widehat{S}_{i\,2^m}|/2^m \leq \varepsilon.$$

This proves (2.5), by (2.6), (2.7) and the arbitrariness of ε, as desired.

Proof of Theorem 1.2. We first show the "if" part. Without loss of generality, assume $t_0 = 2$. Put

$$\widetilde{X}_i = X_i I\{|X_i| \leq (\log i)^{1/\theta}\} - EX_i I\{|X_i| \leq (\log i)^{1/\theta}\}, \quad \widetilde{S}_n = \sum_{i=1}^{\infty} \widetilde{X}_i.$$

Clearly, $Ee^{2|X|^\theta} < \infty$ implies

$$P(|X_n| > (\log n)^{1/\theta} \ i.o.) = 0$$

and

$$\max_{k \geq 0} \sum_{i=k+1}^{k+n} |EX_i I\{|X_i| \leq i^{1/r}\}| \leq \sum_{i=1}^{n} E|X| I\{|X| > (\log i)^{1/\theta}\} = o(n).$$

Thus, we only need to prove that

$$(2.8) \qquad \limsup_{n \to \infty} \max_{0 \leq k \leq e^{\varepsilon n}\, n^\theta} |\widetilde{S}_{n+k} - \widetilde{S}_k|/n = 0 \quad \text{a.s.}$$

It is plain to see that, for any random variable ξ with mean zero and $|\xi| \leq M$ and for any $\varepsilon > 0$,

$$Ee^{t\xi} \leq \exp\left(\frac{t^2}{2} E\xi^2 e^{t|\xi|} \right) \leq \exp\left(\frac{t^2}{2} E\xi^2 e^{tM^{1-\theta}|\xi|^\theta} \right).$$

Note that

$$|\widetilde{X}_i| \leq 2(\log i)^{1/\theta} \leq n, \quad 1 \leq i \leq n + e^{\varepsilon\, n^\theta}$$

provided that n is sufficiently large. Thus, for any $0 < \eta < 1/2$,

$$P\left(\max_{0 \leq k \leq e^{\varepsilon n}\, n^\theta} |\widetilde{S}_{n+k} - \widetilde{S}_k| > \eta\, n \right)$$

$$\leq \sum_{0 \leq k \leq e^{\varepsilon n}\, n^\theta} P\big(|\widetilde{S}_{n+k} - \widetilde{S}_k| > \eta\, n\big)$$

$$(2.9) \qquad \leq 2 \sum_{0 \leq k \leq e^{\varepsilon n}\, n^\theta} \exp\left(- \eta^2\, n^{\theta-1}\, \eta\, n + (\eta^2\, n^{\theta-1})^2\, n\, EX^2\, e^{\eta^2\, n^{\theta-1}\, n^{1-\theta}\, |X|^\theta} \right)$$

$$\leq 4\, e^{\varepsilon n\, n^\theta} \exp\left(- \eta^3\, n^\theta + (\eta^2\, n^{\theta-1})^2\, n\, EX^2\, e^{|X|^\theta} \right)$$

$$\leq 4\, e^{\varepsilon n\, n^\theta} \exp\left(- \eta^3\, n^\theta + \eta^4\, n^\theta\, n^{-(1-\theta)} EX^2\, e^{|X|^\theta} \right)$$

$$\leq 4\, e^{\varepsilon n\, n^\theta} \exp\left(- \eta^3\, n^\theta/2 \right)$$

$$\leq 4 \exp\left(- \eta^3\, n^\theta/4 \right)$$

for every sufficiently large n. (2.8) now follows from (2.9) and the Borel-Cantelli lemma.

We next prove the "only if" part. From (1.3), we have $E|X| < \infty$. Assume $Ee^{t|X|^\theta} = \infty$ for every $t > 0$. It follows that

$$\sum_{n=1}^{\infty} P\big(|X| \geq m\, (\log n)^{1/\theta}\big) = \infty$$

for every positive integer m. Let $k_0 = 0$ and choose k_m recursively so that $k_m \geq 2^m$ and

$$\sum_{n=1+k_{m-1}}^{k_m} P\big(|X| \geq m\, (\log n)^{1/\theta}\big) \geq 1$$

for $m = 1, 2, \cdots$. Let

$$\eta_n = 1/m \quad \text{for} \quad k_{m-1} < n \leq k_m,$$

$$a_n = \max_{1 \leq i \leq n} (\eta_i\, i)^\theta, \quad \varepsilon_n = a_n/n^\theta.$$

Then

$$(2.10) \qquad \sum_{n=1}^{\infty} P\big(|X| \geq (1/\eta_n)\, (\log n)^{1/\theta}\big) = \infty.$$

Since $\varepsilon_n = o(1)$, $Ee^{a(|X|)} < \infty$ by (1.3) and Lemma 2.1. Thus, we have

$$\infty > \sum_{n=1}^{\infty} P(a(|X|) \geq \log n)$$

$$\geq \sum_{n=1}^{\infty} P\big(\eta_{|X|}|X| \geq (\log n)^{1/\theta}\big)$$

$$\geq \sum_{n=1}^{\infty} P\big(\eta_{|X|}|X| \geq (\log n)^{1/\theta}, |X| \leq n\big)$$

$$\geq \sum_{n=1}^{\infty} P\big(|X| \geq (1/\eta_n)(\log n)^{1/\theta}, |X| \leq n\big)$$

$$\geq \sum_{n=1}^{\infty} P\big(|X| \geq (1/\eta_n)(\log n)^{1/\theta}\big) - \sum_{n=1}^{\infty} P(|X| > n).$$

Consequently, we have

$$(2.11) \qquad \sum_{n=1}^{\infty} P\big(|X| \geq (1/\eta_n)(\log n)^{1/\theta}\big) < \infty$$

for $\sum_{n=1}^{\infty} P(|X| > n) \leq E|X| < \infty$, which contradicts (2.10). This shows that there is a $t_0 > 0$ such that $Ee^{t_0|X|^\theta} < \infty$. The proof is now complete.

Proof of Theorem 1.3. Assume that

$$P\Big(\lim_{n \to \infty} (S_{n+k} - S_k)/n = 0 \text{ uniformly in } 0 \leq k \leq e^{c_0 n} \Big) > 0$$

for some $c_0 > 0$. Then, by zero and one law,

$$(2.12) \qquad P\Big(\lim_{n \to \infty} (S_{n+k} - S_k)/n = 0 \text{ uniformly in } 0 \leq k \leq e^{c_0 n} \Big) = 1$$

or, equivalently,

$$(2.13) \qquad \lim_{n \to \infty} \max_{0 \leq k \leq e^{c_0 n}} |S_{n+k} - S_k|/n = 0 \quad \text{a.s.}$$

From Lemma 2.1, it follows that, for any $\varepsilon > 0$,

$$(2.14) \qquad Ee^{c_0|X|/\varepsilon} < \infty.$$

In particular, we have $Ee^{|X|} < \infty$. Set

$$\rho(x) = \inf_t e^{-tx} Ee^{tX}. \quad \alpha_0 = \sup\{x : \rho(x) \ge e^{-c_0}\}.$$

Since $\rho(0) = 1$ and $\rho(x)$ is a strictly decreasing and continuous function in the interval where $\rho(x) > 0$ (cf. Chernoff (1952) or Csörgő and Révész (1981)), $\alpha_0 > 0$. Following the proof of Erdős-Rényi law of large numbers (cf. Csörgő and Révész (1981)), one has

$$(2.15) \qquad \lim_{n\to\infty} \max_{0 \le k \le e^{c_0} n} (S_{n+k} - S_k)/n = \alpha_0 > 0 \quad \text{a.s.}$$

which contradicts (2.13). This proves (1.4).

Along the same lines of the above proof, Theorems 1.1 and 1.2 can be generalized in the following way:

Theorem 2.1. *Let $c > 0$, $1 \le p < 2$, and let $a(x) > 0$ be a nondecreasing function satisfying $a(2x) = O(a(x))$ as $x \to \infty$. Then*

$$P\left(\lim_{n\to\infty} (S_{n+k} - S_k)/n^{1/p} = 0 \ \text{uniformly in } 0 \le k \le c\,a(n)\right) = 1$$

if and only if $EX = 0$ and $E(|X|^p + a(|X|^p)) < \infty$.

Theorem 2.2. *Let $1 \le p < 2$ and $0 < \theta \le 1/p$. Then*

$$P\left(\lim_{n\to\infty} (S_{n+k} - S_k)/n^{1/p} = \mu \ \text{uniformly in } 0 \le k \le e^{\varepsilon_n n^\theta}\right) = 1$$

for any positive sequence $\{\varepsilon_n, n \ge 1\}$ with $\varepsilon_n = o(1)$ if and only if $EX = 0$ and $Ee^{t_0 |X|^{p\theta}} < \infty$ for some $t_0 > 0$.

Acknowledgments. The author wishes to thank Dr. Yeneng Sun for many useful discussions.

REFERENCES

1. H. Chernoff, *A measure of asymptotic efficiency for tests of a hypothesis based on sums of observations*, Ann. Math. Statist. **23** (1952), 493–507.
2. M. Csörgő and P. Révész, *Strong Approximations in Probability and Statistics*, Academic Press, New York, 1981.
3. H. P. Rosenthal, *On the subspace of L^p ($p > 2$) spanned by sequence of independent random variables*, Israel J. Math. **8** (1970), 273–303.
4. Y. Sun, *On metric theorems in the theory of uniform distribution*, Compositio Math. **86** (1993), 15–21.

ON THE PROBABILISTIC STRUCTURE SPACES

YONGFU SU[1]

ABSTRACT. In the present paper, the probabilistic structure space, probabilistic topology spaces, S-type probabilistic metric space, S-type probabilistic normed space and S-type probabilistic inner product space are defined and some relations between these spaces are discussed. Further, we give also the revision of probabilistic $*$-inner product spaces given in [1].

1. Probabilistic Structure Spaces

Let Ω be a nonempty set. A *sigma field* on Ω is a family $\mathcal{A}$ of subsets of Ω such that

(1) $\Omega \in \mathcal{A}$,

(2) $\Omega - A \in \mathcal{A}$ for any $A \in \mathcal{A}$,

(3) if $A_n \in \mathcal{A}$ for $n = 1, 2, \cdots$, then $\cup_{n=1}^{\infty} A_n \in \mathcal{A}$.

Then $(\Omega, \mathcal{A})$ is said to be a *measurable space*.

Let $\mathcal{A}_i$ be sigma field on a set Ω_i $(i = 1, 2)$. A mapping f from Ω_1 into Ω_2 is said to be *measurable* with respect to $\mathcal{A}_i$ $(i = 1, 2)$ if, for all $A_2 \in \mathcal{A}_2$, the inverse $f^{-1}(A_2)$ of A_2 is in $\mathcal{A}_1$.

A *probabilistic measure space* is a triple $(\Omega, \mathcal{A}, P)$, where Ω is a nonempty set, $\mathcal{A}$ is a sigma field on Ω and P is a function from $\mathcal{A}$ into $I = [0, 1]$ satisfying the following conditions:

(1) $P(\Omega) = 1$ and $P(\emptyset) = 0$,

(2) if $\{A_n\}$ is a sequence of pairwise disjoint sets in $\mathcal{A}$, then

$$P(\cup_{n=1}^{\infty} A_n) = \sum_{n=1}^{\infty} P(A_n).$$

Received December 11, 2000.

1991 *Mathematics Subject Classification.* Primary 50A30, 60E05; Secondary 60B05, 60B99.

Key Words and Phrases. Probabilistic structure space, S-type probabilistic metric space, probabilistic metric space, orbit space, probabilistic $*$-inner product space.

[1] Department of Mathematics, Cangzhou Normal College, Cangzhou 061001, People's Republic of China. Email: suyfu@sohu.com

The function P is called a *probabilistic measure* on Ω and a probabilistic measure space $(\Omega, \mathcal{A}, P)$ is said to be *complete* if P is complete, i.e., for $A \in \mathcal{A}$ with $P(A) = 0$, any subset of A is also in $\mathcal{A}$. For more details on probabilistic measure spaces, refer to [3].

Definition 1.1. Let E, S and Ω be nonempty sets, $(S, \bar{\mathcal{A}})$ be a measurable space, (Ω, A, P) be a completely probabilistic measure space and $f : \Omega \to S$ be a measureable mapping with respect to $\mathcal{A}$ and $\bar{\mathcal{A}}$. Then the quadruplet (E, S, Ω, f) is called a *probabilistic structure space* on E.

Definition 1.2. Let (E, S, Ω, f) be a probabilistic structure space.
(1) If S is a family of topologies on E, then (E, S, Ω, f) is called a *probabilistic toplogy space*;
(2) if S is a family of metrics on E, then (E, S, Ω, f) is called a *S-type probabilistic metric space* (shortly, *SPM* space);
(3) if E is a linear space and S is a family of norms on E, then (E, S, Ω, f) is called a *S-type probabilistic normed space* (shortly, SPN-space);
(4) if S is a family of inner products on E, then (E, S, Ω, f) is called a *S-type probabilistic inner product space* (shortly, SPI-space).

2. SPM-Spaces and Probabilistic Metric Spaces

Definition 2.1. Let (E, S, Ω, f) be a SPM-space. If, for every $x, y \in E$ and all $t \in R$,

$$\{d \in S : d(x, y) < t\} \in \bar{\mathcal{A}},$$

then (E, S, Ω, f) is called a *normal SPM-space*.

Theorem 2.1. *Let (E, S, Ω, f) be a normal SPM-space. For every $x, y \in E$ and $t \in R$, define*

$$F_{x,y}(t) = P\{f^{-1}\{d \in S : d(x, y) < t\}\}.$$

Then $(E, F_{x,y}(t), \Delta)$ is an Menger probabilistic metric space, where $\Delta(a, b) = \max\{a + b - 1, 0\}$.

Proof. It is easily seen that $F_{x,y}(t)$ is a distribution function which satisfies the following conditions:
(i) $F_{x,y}(0) = 0$;
(ii) $F_{x,y}(t) = F_{y,x}(t)$;
(iii) $F_{x,y}(t) = H(t)$ if and only if $x = y$,
where

$$H(t) = \begin{cases} 0, & t \leq 0, \\ 1, & t > 0. \end{cases}$$

Next we shall prove that the Menger triangle inequality holds. Indeed, for every $x, y, z \in E$ and $t_1, t_2 \in \mathbb{R}$, we have

$$
\begin{aligned}
F_{x,y}(t_1 + t_2) &= P\{f^{-1}(\{d \in S : d(x,y) < (t_1 + t_2)\})\} \\
&\geq P\{f^{-1}(\{d \in S : d(x,z) < t_1, d(z,y) < t_2\})\} \\
&= P\{f^{-1}(\{d \in S : d(x,z) < t_1\} \cap \{d \in S : d(z,y) < t_2\})\} \\
&= P\{f^{-1}(\{d \in S : d(x,z) < t_1\} + \{d \in S : d(y,z) < t_2\})\} \\
&\quad - P\{f^{-1}(\{d \in S : d(x,z) < t_1\} \cup \{d \in S : d(y,z) < t_2\})\} \\
&\geq F_{x,z}(t_1) + F_{z,y}(t_2) - 1,
\end{aligned}
$$

which implies that

$$
F_{x,y}(t_1 + t_2) \geq \Delta(F_{x,z}(t_1), F_{z,y}(t_2)),
$$

and hence, $(E, F_{x,y}(t), \Delta)$ is a Menger probabilistic metric space. This completes the proof of Theorem 2.1.

Definition 2.2. Let (E, S, Ω, f) be a normal SPM-space and, for $x, y \in E$ and all $t \in R$, let

$$
F_{x,y}(t) = P\{f^{-1}(\{d \in S : d(x,y) < t\})\}.
$$

If $(E, F_{x,y}(t), \Delta)$ is a Menger probabilistic metric space, then it is called an *orbit space* of (E, S, Ω, f).

Definition 2.3. Let $(E, F_{x,y}(t), \Delta)$ be a probabilistic metric space. If, for all $x, y \in E$ with $x \neq y$, $\sup\{t \in R : F_{x,y}(t) = 0\} = 0$, then x and y are said to be *probabilistically equal.*

Theorem 2.2. *Let $(E, F_{x,y}, \min)$ be a Menger probabilistic metric space that does not admit probabilistically equal elements. Then there is a normal SPM-space (E, S, Ω, f) such that $(E, F_{x,y}(t), \min)$ is an orbit space of (E, S, Ω, f).*

Proof. For every $p \in [0, 1)$ and $x, y \in E$, we define a mapping $d_p : E \times E \to R$ as follows:

$$
d_p(x, y) = \sup\{t \in R : F_{x,y}(t) \leq p\}.
$$

Now, we shall prove that d_p is a metric on E.

In fact, $d_p(x, y) \geq 0$ and $d_p(x, y) = d_p(y, x)$ are obvious. Since E has no probabilistically equal elements, we see that $d_p(x, y) = 0$ if and only if $x = y$. If the triangle inequality does not hold, then there are $x, y, z \in E$ and $\epsilon > 0$ such that

$$
d_p(x, y) = d_p(x, z) + d_p(z, y) + 2\epsilon.
$$

It follows that

$$
\begin{aligned}
p &= F_{x,y}(d_p(x,y)) \\
&= F_{x,y}(d_p(x,z) + d_p(z,y) + 2\epsilon) \\
&\geq \min\{F_{x,z}(d_p(x,z) + \epsilon), F_{z,y}(d_p(z,y) + \epsilon)\} \\
&> \min\{p,p\} = p,
\end{aligned}
$$

which is a contradiction.

Let $S = \{d_p : p \in [0,1)\}$ and $\Omega = [0,1)$ be a Lebesgue probabilistic measure space. Let $f : \Omega \to S$ be a mapping such that $f(p) = d_p$ for every $p \in \Omega$. Since f is one to one and f induces a mecause on S, (E, S, Ω, f) is a SPM-space. For every $x, y \in E$ and $t \in R$, we have

$$
\{d_p \in S : d_p(x,y) < t\} = \{d_p \in S : F_{x,y}(t) > p\},
$$

which implies that $\{d_p \in S : d_p(x,y) < t\}$ is a measureable set and so (E, S, Ω, f) is a normal SPM-space.

Finally, we prove that $(E, F_{x,y}(t), \min)$ is an orbit space of (E, S, Ω, f). Indeed,

$$
\begin{aligned}
F_{x,y}(t) &= P\{f^{-1}(\{d_p \in S : F_{x,y}(t) > p\})\} \\
&= P\{f^{-1}(\{d_p \in S : d_p(x,y) < t\})\},
\end{aligned}
$$

where P is a Lebesgue measure on $\Omega = [0,1)$. This completes the proof.

Definition 2.4. Let (E, S, Ω, f) be a SPM space and A be a subset of E. Let

$$
a = P\{f^{-1}(\{d \in S : A \text{ is bounded on } d\})\}.
$$

(1) If $a = 1$, then A is called a *bounded set*;
(2) If $a = 0$, then A is called an *unbounded set*;
(3) If $0 < a < 1$, then A is called a *semi-bounded set*.

Definition 2.5. Let (E, S, Ω, f) be a SPM-space, $\{x_n\}$ be a sequence in E and $x_0 \in E$.

(1) If $P\{f^{-1}(\{d \in S : \lim_{n \to \infty} d(x_n, x_0)\})\} = 1$, then the sequence $\{x_n\}$ is said to be *convergent* to x_0 in probabilistical sense or almost everywhere (shortly, $x_n \to x_0$ a.e.);

(2) If, for any $\epsilon > 0$,

$$
\lim_{n \to \infty} P\{f^{-1}(\{d \in S : d(x_n, x_0) \geq \epsilon\})\} = 0,
$$

then the sequence $\{x_n\}$ is said to be *convergent* to x_0 in measure.

Theorem 2.3. *Let (E, S, Ω, f) be a normal SPM-space. Then*
(1) $x_n \to x_0$ a.e. implies that $x_n \to x_0$ in measure.
(2) $x_n \to x_0$ in measure if and only if $\{x_n\}$ converges to x_0 in probabilistic metric on orbit space.

Proof. (1) Let $\{x_n\}$ be sequences in E and $x_0 \in E$. Then $\{d(x_n, x_0)\}$ is a random variable sequence. It follows from Definition 2.5 that the sequence $\{x_n\}$ converges to x_0 a.e. if and only if the random varibale sequence $\{d(x_n, x_0)\}$ converges to 0 in probabilistic sense and the sequence $\{x_n\}$ converges to x_0 in measure if and only if the sequence $\{d(x_n, x_0)\}$ converges to 0 in measure. Using the probabilistic theory, we know that the convergence in probabilistic implies the convergence in measure. Hence we have (1).
(2) Observe that

$$F_{x_n, x_0}(t) = P\{f^{-1}(\{d \in S : d(x_n, x_0) < t\})\}$$
$$= 1 - P\{f^{-1}(\{d \in S : d(x_n, x_0) \geq t\})\}.$$

It follows that the sequence $\{x_n\}$ converges to x_0 in measure if and only if

$$\lim_{n \to \infty} P\{f^{-1}(\{d \in S : d(x_n, x_0) \geq t\})\} = 0$$

and that the sequence $\{x_n\}$ converges to x_0 in probabilistic metric on orbit space if and only if $\lim_{n \to \infty} F_{x_n, x_0}(t) = 1$ for all $t > 0$. Hence we have (2). This completes the proof.

Theorem 2.4. *Let (E, S, Ω, f) be a normal SPM space and $(E, F_{x,y}(t), \Delta)$ be an orbit space of (E, S, Ω, f). Then*
(i) *A is bounded in (E, S, Ω, f) if and only if A is bounded in $(E, F_{x,y}(t), \Delta)$.*
(ii) *A is unbounded in (E, S, Ω, f) if and only if A is unbounded in $(E, F_{x,y}t), \Delta)$.*
(iii) *A is semi-bounded in (E, S, Ω, f) if and only if A is semi-bounded in $(E, F_{x,y}(t), \Delta)$.*

Proof. Observe that

$$\sup_{t>0} \inf_{x,y \in A} F_{x,y}(t) = \sup_{t>0} \inf_{x,y \in A} P\{f^{-1}(\{d \in S : d(x, y) < t\})\}$$
$$= \sup_{t>0} P\{f^{-1}(\{d \in S : \sup_{x,y \in A} d(x, y) < t\})\}$$
$$= P\{f^{-1}(\{d \in S : \sup_{x,y \in A} d(x, y) < +\infty\})\}.$$

Thus the conclusions (i), (ii) and (iii) are true. This completes the proof.

3. Probabilistic *-Inner Product Spaces

In this section, we revise the definition of probabilistic *-inner product space given in [1].

A triple $(X, F, *)$ is called a *probabilistic *-inner product space* (shortly, a *-PIP-space) ([1]) if X is a real linear space, F is a mapping from $X \times X$ into D satisfying th following conditions, where we denote $F(x, y)$ by $F_{x,y}(t)$ will present the value of $F_{x,y}$ at $t \in \mathbb{R}$:

(*-PI-1) $F_{x,y}(0) = 0$;
(*-PI-2) $F_{x,y} = F_{y,x}$;
(*-PI-3) $F_{x,y}(t) = H(t)$ for all $t \in \mathbb{R}$ if and only if $x = 0$;
(*-PI-4) for all $x, y \in X$ and $t \in \mathbb{R}$,

$$
F_{\alpha x, y}(t) = \begin{cases} F_{x,y}\left(\frac{t}{\alpha}\right) & \text{if } \alpha > 0, \\ H(t) & \text{if } \alpha = 0, \\ 1 - F_{x,y}\left(\frac{t}{\alpha}+\right) & \text{if } \alpha < 0, \end{cases}
$$

where $F_{x,y}\left(\frac{t}{\alpha}+\right)$ is the right limit of Fx, y at $\frac{t}{\alpha}$;
(*-PI-5)

$$
F_{x+y,z}(t) = (F_{x,z} * F_{y,z})(t)
$$

and

$$
(F * G)(t) = \int_{-\infty}^{+\infty} F(t - u)dG(u)
$$

for all $x, y, z \in X$ and $t \in \mathbb{R}$.

We point out the conditions (*-PI-4) and (*-PI-5) are incompatible. To see this, for all $x, y \in X$, define the function

$$
F_{x,y}(t) = \begin{cases} 0 & \text{if } t \leq 0, \\ t & \text{if } 0 < t \leq 0, \\ 1 & \text{if } 1 < t. \end{cases}
$$

By (*-PI-4), we have

$$
F_{x+x,y}(t) = F_{2x,y}(t) = F_{x,y}\left(\frac{t}{2}\right) = \begin{cases} 0 & \text{if } t \leq 0, \\ \frac{t}{2} & \text{if } 0 < t \leq 2, \\ 1 & \text{if } 2 > t. \end{cases}
$$

On the other hand, from ($*$PI5), it follows that

$$F_{x+x,y}(t) = \int_{-\infty}^{+\infty} F_{x,y}(t-u)dF_{x,y}(u)$$

$$= \int_{-\infty}^{0} F_{x,y}(t-u)dF_{x,y}(u) + \int_{0}^{1} F_{x,y}(t-u)dF_{x,y}(u)$$

$$+ \int_{1}^{+\infty} F_{x,y}(t-u)dF_{x,y}(u)$$

$$= \int_{0}^{1} F_{x,y}(t-u)dF_{x,y}(u) = \int_{0}^{1} F_{x,y}(t-u)d(u)$$

$$= - \int_{0}^{1} F_{x,y}(t-u)d(t-u) = - \int_{t}^{t-1} F_{x,y}(u)d(u)$$

$$= \int_{t}^{t-1} F_{x,y}(u)d(u),$$

which yields that

$$F_{x+x,y}(t) = \begin{cases} 0 & \text{if } t \le 0, \\ \frac{t^2}{2} & \text{if } 0 < t \le 1, \\ \frac{-t^2}{2} + 2t - 1 & \text{if } 1 < t \le 2, \\ 1 & \text{if } 2 < t. \end{cases}$$

This is a contradiction.

Thus, we give another definition of probabilistic $*$-inner product space, which revises the corresponding definition of the reference [1].

A triple $(X, F, *)$ is called a *probabilistic $*$-inner product space* (shortly, a $*$-PIP-space) if X is a real linear space, F is a mapping from $X \times X$ into D satisfying the following conditions, where we denote $F(x,y)$ by $F_{x,y}(t)$ will present the value of $F_{x,y}$ at $t \in \mathbb{R}$:

($*$-PI-1) $F_{x,y}(0) = 0$;

($*$-PI-2) $F_{x,y} = F_{y,x}$;

($*$-PI-3) $F_{x,y}(t) = H(t)$ for all $t \in \mathbb{R}$ if and only if $x = 0$;

($*$-PI-4) for all $x, y \in \mathbb{R}$ and $t \in \mathbb{R}$,

$$F_{x,y}(t) = \begin{cases} F_{x,y}\left(\frac{t}{\alpha}\right) & \text{if } \alpha > 0, \\ H(t) & \text{if } \alpha = 0, \\ 1 - F_{x,y}\left(\frac{t}{\alpha}+\right) & \text{if } \alpha < 0, \end{cases}$$

where $F_{x,y}\left(\frac{t}{\alpha}+\right)$ is the right limit of Fx, y at $\frac{t}{\alpha}$;
($*$-PI-5) for all $x, y, z \in X$ and $t \in \mathbb{R}$,

$$F_{x+y,z}(t) = (F_{x,z} * F_{y,z})(t)$$

whenever x and y are linearly independent.

Remark. The definition above revises the corresponding definition of the reference [1] and the relevant results of the reference [1] are still true under our definition of this paper.

REFERENCES

1. Shin-Sen Chang, Yeol Je Cho and Shin Min Kang, *Probabilistic Metric Spaces and Nonliear Operator theory*, Sichuan Uiversity Press, 1994.
2. B. Schweieer, *The Metrization statistical space*, Pacific J. Math. **10** (1960), 673–615.
3. B. Schloeizer and A. Sklar, *Probabilistic Metric Spaces*, North-Horland, 1983.

LIMIT BEHAVIORS FOR THE INCREMENTS
OF A d-DIMENSIONAL GAUSSIAN PROCESS

HWA SANG SUNG[1] AND YONG KAB CHOI[1]

ABSTRACT. In this paper, we study limit behaviors for the increments of a d-dimensional Gaussian process with independent copies, via estimating upper bounds of large deviation probabilities on the suprema of the d−dimensional Gaussian process.

1. Introduction and Results

The limit theorems on increments of Wiener processes and Gaussian processes have been investigated in various directions by many authors, for instance, Csáki, Csörgő and Shao [4], Csörgő and Shao [6], Csörgő and Révész [5], Ortega [10], Ortega and Wschebor [11], Choi [1], Csáki et al. [3], Lin and Lu [9], Zhang [13] and etc.

In this paper, we are interested in studying limit behaviors for the increments of a d-dimensional Gaussian process. Throughout the paper we shall always assume the following statements: Let $\{X(t),\ t \in [0,\infty)\}$ be a real-valued continuous, centered Gaussian process with $X(0) = 0$ and stationary increment

$$E\{X(t) - X(s)\}^2 = \sigma^2(|t - s|),$$

where $\sigma(t),\ t > 0$, is a positive nondecreasing continuous and regularly varying function with exponent $\alpha\,(0 < \alpha < 1)$ at 0 and ∞. Let $\{X^d(t) = (X_1(t),\cdots,$ $X_d(t)) \in R^d, t \in [0,\infty)\}$ be a d-dimensional Gaussian process with copies $X_i(t)$ $(i = 1,\cdots,d)$ of $X(t)$.

Received November 9, 2000.

1991 *Mathematics Subject Classification*. 60F15, 60G15.

Key Words and Phrases. Gaussian process, Borel-Cantelli lemma, regularly varying function.

This work was supported by grant No. 2001-1-10400-010-2 from the Basic Research Program of the Korea Science & Engineering Foundation.

[1] Department of Mathematics, Gyeongsang National University, Chinju 660-701, Korea. Email: sungyhg@hanmail.net and mathykc@nongae.gsnu.ac.kr

For $0 < T < \infty$, let a_T and b_T be real-valued positive functions of T such that

(i)
$$\liminf_{T \to \infty} a_T > 0.$$

For convenience, we denote:

$$\beta_1(T) = \left\{ 2\left(\log \frac{b_T}{a_T} + \log\log\left(\sigma(a_T) + \frac{1}{\sigma(a_T)} \right) \right) \right\}^{1/2},$$

$$\beta_2(T) = \left\{ 2\left(\log \frac{b_T}{a_T} + \log\log\left(a_T + \frac{1}{a_T} \right) \right) \right\}^{1/2},$$

where $\log x = \ln(x \vee 1)$, $a \vee b = \max\{a, b\}$. Note that $\beta_1(T) \le \beta_2(T)$ for all large T.

The main results are as follows:

Theorem 1.1. *Assume that*

(ii)
$$\frac{b_T}{a_T} + a_T + \frac{1}{a_T} \to \infty \quad \text{as } T \to \infty.$$

Then we have

(1.4)
$$\limsup_{T \to \infty} \ \sup_{0 \le t \le b_T} \ \sup_{0 \le s \le a_T} \ \frac{\|X^d(t+s) - X^d(t)\|}{\sigma(a_T)\,\beta_1(T)} \le 1 \quad a.s.$$

Theorem 1.2. *Assume that*

(iii)
$$\frac{\log(b_T/a_T)}{\log\log\left(a_T + \frac{1}{a_T} \right)} \to \infty \quad \text{as } T \to \infty$$

and there exists a constant $c > 0$ such that, for $h > 0$,

(iv)
$$\left| \frac{d^2\sigma^2(h)}{dh^2} \right| \le c\,\frac{\sigma^2(h)}{h^2}.$$

Then we have

$$\liminf_{T \to \infty} \ \sup_{0 \le t \le b_T} \ \frac{\|X^d(t+a_T) - X^d(t)\|}{\sigma(a_T)\,\beta_2(T)} \ge 1, \quad a.s.$$

From Theorems 1.1 and 1.2, we obtain the following limit theorem:

Corollary 1.1. *Under the asumptions of Theorem 1.2, we have*

$$\lim_{T\to\infty} \sup_{0\le t\le b_T} \sup_{0\le s\le a_T} \frac{\|X^d(t+s) - X^d(t)\|}{\sigma(a_T)\,\beta_i(T)}$$

$$= \lim_{T\to\infty} \sup_{0\le t\le b_T} \frac{\|X^d(t+a_T) - X^d(t)\|}{\sigma(a_T)\,\beta_i(T)} = 1, \quad i = 1,2 \quad a.s.$$

Example. Let $\{X(t),\ 0 \le t < \infty\}$ be a fractional Brownian motion of order 2α with $0 < \alpha < 1$, that is, let $\{X(t),\ 0 \le t < \infty\}$ be a real-valued Gaussian process with mean 0, $X(0) = 0$ and $\sigma(t) = t^\alpha$. When $\alpha = 1/2$, $\{X(t),\ 0 \le t < \infty\}$ is a standard Wiener process. Let $a_T = 1$ and $b_T = e^T$. Then by Corollary 1.1 we have

$$\lim_{T\to\infty} \sup_{0\le t\le e^T} \sup_{0\le s\le 1} \frac{\|X^d(t+s) - X^d(t)\|}{\sqrt{T}}$$

$$= \lim_{T\to\infty} \sup_{0\le t\le e^T} \frac{\|X^d(t+1) - X^d(t)\|}{\sqrt{T}} = \sqrt{2} \quad a.s.$$

On the other hand, letting $a_T = 1/T$ and $b_T = 1$, we get moduli of continuity

$$\lim_{T\to\infty} \sup_{0\le t\le 1} \sup_{0\le s\le 1/T} \frac{\|X^d(t+s) - X^d(t)\|}{T^{-\alpha}\sqrt{\log T}}$$

$$= \lim_{T\to\infty} \sup_{0\le t\le 1} \frac{\|X^d(t+1/T) - X^d(t)\|}{T^{-\alpha}\sqrt{\log T}} = \sqrt{2} \quad a.s.$$

2. The Proofs

The following Lemmas 2.1 and 2.2 are essential to prove Theorem 1.1. Let $\mathbb{D}$ be a compact subset of R^N with the usual Euclidean norm $\|\cdot\|$ and let $\{X(\mathbf{t}),\ \mathbf{t} \in \mathbb{D}\}$ be a real-valued separable centered Gaussian process with $E\{X(\mathbf{t}) - X(\mathbf{s})\}^2 = \sigma^2(\|\mathbf{t} - \mathbf{s}\|)$. Suppose that

$$(2.1) \qquad\qquad 0 < \Gamma := \sup_{\mathbf{t}\in\mathbb{D}} \big\{ E(X(\mathbf{t}))^2 \big\}^{1/2} < \infty,$$

$$(2.2) \qquad\qquad \sigma(h) \le \varphi(h),$$

where $\varphi(\cdot)$ is a nondecreasing continuous function. Let $\{X^d(\mathbf{t}) = (X_1(\mathbf{t}), \cdots, X_d(\mathbf{t})) \in R^d,\ \mathbf{t} \in \mathbb{D}\}$ be a d-dimensional Gaussian process whose components $X_i(\mathbf{t}),\ i = 1, \cdots, d$, are copies of $X(\mathbf{t})$. Denote the Lebesgue measure of $\mathbb{D}$ by $m(\mathbb{D})$.

The following Lemma 2.1 is another version of Fernique's inequality [7] on the d-dimensional Gaussian process, whose proof is similar to that of Lemma 2.3 in [2].

Lemma 2.1. *For $\lambda > 0, x \geq 1$ and $B > (2\sqrt{2}+2)\sqrt{2N\log 2}$, we have*

$$P\left\{\sup_{t\in\mathbb{D}}\|X^d(t)\| > x\Big(\Gamma + B\int_0^\infty \varphi(\sqrt{N}\lambda 2^{-y^2})\,dy\Big)\right\} \leq C\frac{m(\mathbb{D})}{\lambda^N}\Phi_d(x),$$

where $C > 0$ is a constant, $\Phi_d(x) = P\{\|N^d(0,1)\| \geq x\}$ and $N^d(0,1)$ denotes a d-dimensional standardized normal random vector.

Using Lemma 2.1, we can obtain an upper bound for the following large deviation probabilities by the same lines as the proof of Lemma 2.4 in [2]:

Lemma 2.2. *Let a_T and b_T be positive functions of $T > 0$. Then, for any $\varepsilon > 0$, there exists a positive constant C_ε depending only on ε such that, for all $u > 0$,*

$$P\left\{\sup_{0\leq t\leq b_T}\sup_{0\leq s\leq a_T}\frac{\|X^d(t+s)-X^d(t)\|}{\sigma(a_T)} \geq u\right\}$$
$$\leq C_\varepsilon\frac{b_T}{a_T}u^{d-2}\exp\left(-\frac{u^2}{2+\varepsilon}\right).$$

Proof of Theorem 1.1. Let $\theta = \sqrt{1+\varepsilon}$ for any given $\varepsilon > 0$. Define

$$A_k = \{T : \theta^k \leq \sigma(a_T) \leq \theta^{k+1}\}, \quad -\infty < k < \infty,$$

$$A_{k,j} = \{T : 2^j \leq \frac{b_T}{a_T} \leq 2^{j+1}, \ T \in A_k\}, \quad j \geq 0,$$

$$a_{T_{k,j}} = \sup\{a_T : T \in A_{k,j}\}, \quad b_{T_{k,j}} = \sup\{b_T : T \in A_{k,j}\}.$$

Considering the condition (ii), we have

$$\limsup_{T\to\infty}\sup_{0\leq t\leq b_T}\sup_{0\leq s\leq a_T}\frac{\|X^d(t+s)-X^d(t)\|}{\sigma(a_T)\beta_1(T)}$$

$$(2.8)\qquad \leq \limsup_{|k|+l\to\infty}\sup_{j\geq l}\sup_{T\in A_{k,j}}\sup_{0\leq t\leq b_T}\sup_{0\leq s\leq a_T}\frac{\|X^d(t+s)-X^d(t)\|}{\sigma(a_T)\beta_1(T)}$$

$$\leq \limsup_{|k|+l\to\infty}\sup_{j\geq l}\sup_{0\leq t\leq b_{T_{k,j}}}\sup_{0\leq s\leq a_{T_{k,j}}}\frac{\|X^d(t+s)-X^d(t)\|}{\theta^k\big\{2\big(\log 2^j + \log\log\theta^{|k|}\big)\big\}^{1/2}}.$$

We will show that

$$(2.9)\qquad \limsup_{|k|+l\to\infty}\sup_{j\geq l}\sup_{0\leq t\leq b_{T_{k,j}}}\sup_{0\leq s\leq a_{T_{k,j}}}\frac{\|X^d(t+s)-X^d(t)\|}{\theta^k\big\{2\big(\log 2^j + \log\log\theta^{|k|}\big)\big\}^{1/2}}$$

$$\leq \theta\limsup_{|k|+l\to\infty}\sup_{j\geq l}\sup_{0\leq t\leq b_{T_{k,j}}}\sup_{0\leq s\leq a_{T_{k,j}}}\frac{\|X^d(t+s)-X^d(t)\|}{\sigma(a_{T_{k,j}})\beta_1(k,j)} \leq \theta^2 \quad \text{a.s.}$$

where $\beta_1(k,j) = \{2(\log 2^j + \log\log\theta^{|k|})\}^{1/2}$. Let $k \neq 0$. Then, by Lemma 2.2, there exists $C_\varepsilon > 0$ such that

$$
P\left\{\sup_{j\geq l}\ \sup_{0\leq t\leq b_{T_{k,j}}}\ \sup_{0\leq s\leq a_{T_{k,j}}} \frac{\|X^d(t+s) - X^d(t)\|}{\sigma(a_{T_{k,j}})\,\beta_1(k,j)} \geq \theta\right\}
$$

$$
\leq \sum_{j\geq l} P\left\{\sup_{0\leq t\leq b_{T_{k,j}}}\ \sup_{0\leq s\leq a_{T_{k,j}}} \frac{\|X^d(t+s) - X^d(t)\|}{\sigma(a_{T_{k,j}})} \geq \theta\,\beta_1(k,j)\right\}
$$

$$
\text{(2.10)} \qquad \leq C_\varepsilon \sum_{j\geq l}\left(\frac{b_{T_{k,j}}}{a_{T_{k,j}}}\right)(\theta\cdot\beta_1(k,j))^{d-2}\ \exp\left(-\frac{1}{2+\varepsilon}\theta^2\,\beta_1^2(k,j)\right)
$$

$$
\leq C_\varepsilon \sum_{j\geq l}\left(\frac{b_{T_{k,j}}}{a_{T_{k,j}}}\right)\ \exp\left(-\frac{2}{2+\varepsilon}\,\theta^2\left(\log 2^j + \log\log\theta^{|k|}\right)\right)
$$

$$
\leq C_\varepsilon \sum_{j\geq l}(2^j)^{1-(2\theta^2)/(2+\varepsilon)}\,|k|^{-2\theta^2/(2+\varepsilon)}
$$

$$
\leq C_\varepsilon\,|k|^{-1-\delta}\,2^{-\delta l},
$$

where $\delta = \varepsilon/(2+\varepsilon)$. Hence we have

$$
\text{(2.11)} \qquad \sum_{l=0}^{\infty}\sum_{|k|=1}^{\infty} P\left\{\sup_{j\geq l}\ \sup_{0\leq t\leq b_{T_{k,j}}}\ \sup_{0\leq s\leq a_{T_{k,j}}} \frac{\|X^d(t+s) - X^d(t)\|}{\sigma(a_{T_{k,j}})\,\beta_1(k,j)} \geq \theta\right\} < \infty.
$$

If $k = 0$, then, by the same way as $(2.8)\sim(2.10)$, we have

$$
\text{(2.12)} \qquad
\begin{aligned}
&P\left\{\sup_{j\geq l}\ \sup_{0\leq t\leq b_{T_{k,j}}}\ \sup_{0\leq s\leq a_{T_{k,j}}} \frac{\|X^d(t+s) - X^d(t)\|}{\sigma(a_{T_{k,j}})\,\beta_1(0,j)} \geq \theta\right\} \\
&\qquad \leq C_\varepsilon \sum_{j\geq l}\frac{b_{T_{k,j}}}{a_{T_{k,j}}}\exp\left(-\frac{\theta^2}{2+\varepsilon}(2\log 2^j)\right) \leq C_\varepsilon 2^{-\delta l}
\end{aligned}
$$

and

$$
\text{(2.13)} \qquad \sum_{l=0}^{\infty} P\left\{\sup_{j\geq l}\ \sup_{0\leq t\leq b_{T_{k,j}}}\ \sup_{0\leq s\leq a_{T_{k,j}}} \frac{\|X^d(t+s) - X^d(t)\|}{\sigma(a_{T_{k,j}})\beta_1(0,j)} \geq \theta\right\} < \infty.
$$

Applying the Borel-Cantelli lemma to (2.11) and (2.13) yields (2.9) and hence we obtain (1.4) from (2.8) since θ is arbitrary.

To prove Theorem 1.2, we need the following Lemmas 2.3 and 2.4:

Lemma 2.3. [8] *Let $\{\xi_l : l = 1, 2, \cdots, n\}$ be jointly standardized normal random variables with $\Lambda_{ll'} = Corr(\xi_l, \xi_{l'})$ such that $\delta := \max_{l \neq l'} |\Lambda_{ll'}| < 1$. Then for any real number u and integers $1 \leq m_1 < m_2 < \cdots < m_k \leq n$ with $k \leq n$,*

$$(2.14) \quad P\left\{ \max_{1 \leq l \leq k} \xi_{m_l} \leq u \right\} \leq \{\Phi(u)\}^k + c \sum_{1 \leq l < l' \leq k} |\lambda(l, l')| \exp\left(-\frac{u^2}{1 + |\lambda(l, l')|}\right),$$

where $\lambda(l, l') = \Lambda(m_l, m_{l'})$, and $c = c(\delta)$ is a constant independent of n, u and k, and we denote $\Phi(u) = \int_{-\infty}^{u} \frac{1}{\sqrt{2\pi}} e^{-y^2/2} \, dy$.

Estimating an upper bound for the second term of the right hand side of (2.14), we obtain the following:

Lemma 2.4. [1] *Let ξ_l, δ, k and $\lambda(l, l')$ be as in Lemma 2.3. Further, assume that the inequality $|\lambda(l, l')| < |l - l'|^{-\nu}$ holds for some $\nu > 0$. Set $u = \{(2 - \eta) \log k\}^{1/2}$, where $0 < \eta < (1 - \delta)\nu/(1 + \nu + \delta)$. Then we have*

$$\sum := \sum_{1 \leq l < l' \leq k} |\lambda(l, l')| \exp\left(-\frac{u^2}{1 + |\lambda(l, l')|}\right) \leq ck^{-\delta_0},$$

where $\delta_0 = \{\nu(1 - \delta) - \eta(1 + \delta + \nu)\}/\{(1 + \nu)(1 + \delta)\} > 0$ and c is a positive constant independent of n, u and k.

Proof of Theorem 1.2. For any fixed $\theta > 1$, define

$$B_{k,j} = \{T : \theta^{k-1} \leq b_T \leq \theta^k, \ \theta^{j-1} \leq a_T \leq \theta^j\}, \ k \geq 1, \ -\infty < j < \infty.$$

The condition (iii) implies that there exists $M > 0$ sufficient large such that $b_T/a_T \geq M$ for all large $T > 0$, and let $k(\theta) = [\theta^{k-1-j}/M]$. Then, by (iii), we have

$$\sup_{T \in B_{k,j}} \beta_2(T) \leq \theta\{2 \log k(\theta)\}^{1/2},$$

$$(2.15) \qquad \inf_{T \in B_{k,j}} \beta_2(T) \geq \{2(\log \theta^{k-1-j} + \log \log(\theta^{j-1} + \theta^{-j}))\}^{1/2}$$

$$\geq \theta^{-1}\{2 \log \theta^{k-j}\}^{1/2}$$

for k large enough. From the regular variation of $\sigma(\cdot)$, we have

$$(2.16) \qquad \frac{\sigma(\theta^j - \theta^{j-1})}{\sigma(\theta^{j-1})} \leq (\theta - 1)^{\alpha}$$

for some $0 < \alpha < 1$. By the conditions (i) and (iii), there exists $c < \infty$ such that $-c < j \le k/2$. Hence, by (2.15)

$$
\begin{aligned}
\liminf_{T \to \infty} \sup_{0 \le t \le b_T} & \frac{\|X^d(t + a_T) - X^d(t)\|}{\sigma(a_T)\,\beta_2(T)} \\
& \ge \liminf_{k \to \infty} \inf_{-c < j \le k/2} \inf_{T \in B_{k,j}} \sup_{0 \le t \le b_T} \frac{\|X^d(t + a_T) - X^d(t)\|}{\sigma(a_T)\,\beta_2(T)} \\
& \ge \frac{1}{\theta} \liminf_{k \to \infty} \inf_{-c < j \le k/2} \sup_{0 \le t \le \theta^{k-1}} \frac{\|X^d(t + \theta^j) - X^d(t)\|}{\sigma(\theta^j)\{2 \log k(\theta)\}^{1/2}} \\
& \quad - \theta \limsup_{k \to \infty} \sup_{-c < j \le k/2} \sup_{0 \le t \le \theta^k} \sup_{\theta^{j-1} \le s \le \theta^j} \frac{\|X^d(t + s) - X^d(t + \theta^j)\|}{\sigma(\theta^{j-1})\{2 \log \theta^{k-j}\}^{1/2}}.
\end{aligned}
$$

(2.17)

Let ξ_i, $1 \le i \le d$, be normal random variables with means zero and let a_i, $i = 1, \cdots, d$, be positive numbers. Put $q = p/(p-1)$ for $p > 1$ and define $\|\mathbf{a}\|_q = \left(\sum_{i=1}^{d} a_i^q\right)^{1/q}$ for $\mathbf{a} = (a_1, \cdots, a_d)$. It is well-known that

$$
\left(\sum_{i=1}^{d} |\xi_i|^p\right)^{1/p} = \sup_{\|\mathbf{a}\|_q \le 1} \sum_{i=1}^{d} a_i \xi_i.
$$

(see Lemma 2.3 in [12]). Setting $a_i = 1/\sqrt{d}$ $(i = 1, \cdots, d)$, we have

$$
\frac{\|X^d(t + \theta^j) - X^d(t)\|}{\sigma(\theta^j)} \ge \frac{\sum_{i=1}^{d} \left(X_i(t + \theta^j) - X_i(t)\right)}{\sqrt{d}\,\sigma(\theta^j)}.
$$

Let

$$
\mathbb{Z}_\theta(l) = \frac{\sum_{i=1}^{d} \left(X_i(lM\theta^j + \theta^j) - X_i(lM\theta^j)\right)}{\sqrt{d}\,\sigma(\theta^j)}, \qquad 0 \le l \le k(\theta).
$$

Then $\mathbb{Z}_\theta(l)$ are standard normal random variables. It follows from (2.16)$\sim$(2.17) that the right hand side of (2.17) is greater than or equal to

$$
\begin{aligned}
& \frac{1}{\theta} \liminf_{k \to \infty} \inf_{-c < j \le k/2} \max_{0 \le l \le k(\theta)} \frac{\mathbb{Z}_\theta(l)}{\{2 \log k(\theta)\}^{1/2}} \\
& \quad - \theta(\theta - 1)^\alpha \limsup_{k \to \infty} \sup_{-c < j \le k/2} \sup_{0 \le t \le \theta^k} \sup_{\theta^{j-1} \le s \le \theta^j} \\
& \qquad\qquad \frac{\|X^d(t + s) - X^d(t + \theta^j)\|}{\sigma(\theta^j - \theta^{j-1})\{2 \log \theta^{k-j}\}^{1/2}} \\
& =: J_1 - J_2.
\end{aligned}
$$

(2.18)

Using (iv) and the relation $ab = \left(a^2 + b^2 - (a-b)^2\right)/2$, it follows that, for $l > l'$,

$$\left|\lambda_\theta(l, l')\right| := \left|Corr\big(\mathbb{Z}_\theta(l),\ \mathbb{Z}_\theta(l')\big)\right|$$

$$= \frac{1}{d\,\sigma^2(\theta^j)}\Big|\sum_{i=1}^{d} E\{X_i(lM\theta^j + \theta^j)X_i(l'M\theta^j + \theta^j)$$

$$- X_i(lM\theta^j + \theta^j)X_i(l'M\theta^j) - X_i(lM\theta^j)X_i(l'M\theta^j + \theta^j)$$

$$+ X_i(lM\theta^j)X_i(l'M\theta^j)\}\Big|$$

$$\leq \frac{1}{2\sigma^2(\theta^j)}\Big|\big(\sigma^2((l-l')M\theta^j + \theta^j) - \sigma^2((l-l')M\theta^j)\big)$$

$$- \big(\sigma^2((l-l')M\theta^j) - \sigma^2((l-l')M\theta^j - \theta^j)\big)\Big|$$

$$\leq \frac{1}{2\sigma^2(\theta^j)}\int_{(l-l')M\theta^j - \theta^j}^{(l-l')M\theta^j} \left|\sigma^2(t+\theta^j)' - \sigma^2(\theta^j)'\right| dt$$

$$\leq \frac{1}{2\sigma^2(\theta^j)}\int_{(l-l')M\theta^j - \theta^j}^{(l-l')M\theta^j} \left(\int_t^{t+\theta^j} \left|\sigma^2(y)''\right| dy\right) dt$$

$$\leq \frac{1}{2\sigma^2(\theta^j)}\int_{(l-l')M\theta^j - \theta^j}^{(l-l')M\theta^j} \left(\int_t^{t+\theta^j} c\,\frac{\sigma^2(y)}{y^2}\,dy\right) dt$$

$$\leq \frac{c\,\sigma^2((l-l')M\theta^j + \theta^j)}{\sigma^2(\theta^j)((l-l')M - 1)^2} < |l - l'|^{-\nu},$$

where $\nu = 2 - 2\alpha > 0$, and we used the regular variation of $\sigma(\cdot)$ in the last inequality. Let us now apply Lemmas 2.3 and 2.4 for

$$\xi_l = \mathbb{Z}_\theta(l), \qquad 0 \leq l \leq k(\theta),$$

$$|\lambda(l,\, l')| = |\lambda_\theta(l,\, l')| < |l - l'|^{-\nu},$$

$$u = \{(2 - \eta)\log k(\theta)\}^{1/2}, \qquad \eta = 2\varepsilon < \frac{(1-\delta)\nu}{1 + \nu + \delta}.$$

Then, for any $0 < \epsilon < 1$, we have

$$P\left\{\max_{0 \leq l \leq k(\theta)} \frac{\mathbb{Z}_\theta(l)}{\{2\log k(\theta)\}^{1/2}} \leq \sqrt{1-\epsilon}\right\} \leq \{\Phi(u)\}^{k(\theta)} + c\,k(\theta)^{-\delta_0}$$

$$\leq \exp\left(-c\,k(\theta)^\epsilon\right) + c\,k(\theta)^{-\delta_0} \leq c\,k(\theta)^{-\delta_0}.$$

Thus we obtain

$$P\left\{\inf_{-c < j \leq k/2}\ \max_{0 \leq l \leq k(\theta)} \frac{\mathbb{Z}_\theta(l)}{\{2\log k(\theta)\}^{1/2}} \leq \sqrt{1-\epsilon}\right\}$$

$$\leq \sum_{-c < j \leq k/2} c\,k(\theta)^{-\delta_0} \leq c\,k\theta^{-k\delta_0/2}$$

and the Borel-Cantelli lemma yields that

$$(2.19) \qquad\qquad J_1 \geq 1 \quad a.s.$$

since θ and ϵ are arbitrary. From $(2.17)\sim(2.19)$, the proof is completed if we show that

$$(2.20) \qquad\qquad J_2 \leq \epsilon \quad a.s.$$

for any $\epsilon > 0$. Again by Lemma 2.2, for any $\epsilon > 0$, there exists $C_\varepsilon > 0$ such that for large k

$$P\left\{ \sup_{-c<j\leq k/2} \sup_{0\leq t\leq \theta^k} \sup_{\theta^{j-1}\leq s\leq \theta^j} \frac{\|X^d(t+s)-X^d(t+\theta^j)\|}{\sigma(\theta^j-\theta^{j-1})\{2\log\theta^{k-j}\}^{1/2}} \geq \sqrt{2(1+\varepsilon)} \right\}$$

$$\leq C_\varepsilon \sum_{-c<j\leq k/2} \theta^{k-j} \exp\left(-\frac{4(1+\varepsilon)}{2+\varepsilon}\log\theta^{k-j}\right) \leq C_\varepsilon(\theta^{1/2})^{-k}.$$

Hence we have

$$\sum_{k=1}^{\infty} P\left\{ \sup_{-c<j\leq k/2} \sup_{0\leq t\leq \theta^k} \sup_{\theta^{j-1}\leq s\leq \theta^j} \frac{\|X^d(t+s)-X^d(t+\theta^j)\|}{\sigma(\theta^j-\theta^{j-1})\{2\log\theta^{k-j}\}^{1/2}} \geq \sqrt{2(1+\varepsilon)} \right\}$$

$$< \infty.$$

By the Borel-Cantelli lemma, it follows that

$$(2.21) \qquad \limsup_{k\to\infty} \sup_{-c<j\leq k/2} \sup_{0\leq t\leq \theta^k} \sup_{\theta^{j-1}\leq s\leq \theta^j} \frac{\|X^d(t+s)-X^d(t)\|}{\sigma(\theta^j-\theta^{j-1})\{2\log\theta^{k-j}\}^{1/2}}$$

$$\leq \sqrt{2} \quad a.s.$$

Combining (2.18) with (2.21), we have (2.20) since θ is arbitrary.

REFERENCES

1. Y. K. Choi, *Erdös-Rényi type laws applied to Gaussian processes*, J. Math. Kyoto Univ. **31** (1991), 191–217.

2. Y. K. Choi and S. H. Lee, *Some limit theorems on the increments of l^p-valued Gaussian processes*, Submitted in Stochastic Analysis and Applications, Nova Science Publishers, Inc., New York, 2001.

3. E. Csáki, M. Csörö, Z. Y. Lin and P. Révész, *On infinite series of independent Ornstein-Uhlenbeck processes*, Stoch. Proc. & Their Appl. **39** (1991), 25–44.

4. E. Csáki, M. Csörö and Q. M. Shao, *Fernique type inequalities and modui of continuity for l^2-valued Ornstein-Uhlenbeck processes*, Ann. Inst. H. poincaré **28** (1992), 479–517.

5. M. Csörgő and P. Révész, *Strong approximations in Probability and Statistics*, Academic Press, New York, 1981.

6. M. Csörő and Q. M. Shao, *Strong limit theorems for large and small increments of l^2-valued Gaussian processes*, Ann. Probab. **21** (1993), 1958–1990.

7. X. Fernique, *Continuité des processus Gaussiens*, C. R. Acad. Sci. Paris, t. **258** (1964), 6058–6060.

8. M. R. Leadbetter, G. Lindgren and H. Rootzen, Extremes and Related Properties of Random Sequences and Processes, Springer-Verlag, New York, 1983, pp. 81–84.

9. Z. Y. Lin and C. R. Lu, *Strong limit theorems*, Science Press, Kluwer Academic Publishers, 1992.

10. J. Ortega, *On the size of the increments of non-stationary Gaussian processes*, Stoch. Process Appl. **18** (1984), 47–56.

11. J. Ortega and M. Wschebor, *On the increments of the Wiener process*, Z. Wahrsch. verw. Gebiete **65** (1984), 329–339.

12. Q. M. Shao, *p-variation of Gaussian processes with stationary increments*, Studia Sci. Math. Hungar. **31** (1996), 237–247.

13. L. X. Zhang, *Some liminf results on increments of fractional Brownian motion*, Acta Math. Hungar. **71(3)** (1996), 215–240.

ON LARGE INCREMENTS AND LIMINFS OF
A TWO-PARAMETER GAUSSIAN PROCESS

ZHANG LI-XIN[*†1], LU CHUANRONG[1] AND WANG YAOHUNG[‡2]

ABSTRACT. In this paper, how big the increments are and some liminf behaviors are studied of a two-parameter Gaussian process which is a generalization of a two-parameter Wiener process. The results are based on some inequalities on the suprema of this process which also are of independent interest.

1. Introduction and Main Results.

Let $\{Z(t,s); t,s \geq 0\}$ be a mean zero Gaussian process with $Z(0,0) = 0$ a.s. and

$$EZ(t_1, s_1)Z(t_2, s_2)$$
$$= \{|t_1|^{2\alpha} + |t_2|^{2\alpha} - |t_2 - t_1|^{2\alpha}\}\{|s_1|^{2\alpha} + |s_2|^{2\alpha} - |s_2 - s_1|^{2\alpha}\}/4.$$

It is called the *Brownian-Einsteinian-Smoluchowskian Wiener* (for short BESW) *process* of order α $(0 < \alpha < 1)$ or the two-parameter fractional Wiener process of order α. If $\alpha = 1/2$, then $\{Z(t,s); t,s \geq 0\}$ is a standard two-parameter Wiener process, we denote it $\{W(t,s); t,s \geq 0\}$. The BESW process is different from the two-parameter fractional Lévy Brownian motion or so called fractional Lévy Wiener process which is studied by many authors. A two-parameter fractional Lévy Brownian motion is a mean zero Gaussian process $\{X(t,s); t,s \geq 0\}$ with $X(0,0) = 0$ a.s. and

$$EX(t_1, s_1)X(t_2, s_2)$$
$$= \{|t_1^2 + t_2^2|^{\alpha} + |s_1^2 + s_2^2|^{\alpha} - |(t_1 - t_2)^2 + (s_1 - s_2)^2|^{\alpha}\}/2.$$

Received November 7, 2000.

1991 *Mathematics Subject Classification.* Primary 60F15; Secondary 60G15.

Key Words and Phrases. Fractional Wiener process, increments, liminfs.

†The corresponding author and Research supported by National Natural Science Foundation of China (No. 19701011).

‡Supported by NSC 88-2118-M029-001 of Taiwan.

[1] Department of Mathematics, Zhejiang University, Xixi Campus, Zhejiang, Hangzhou 310028, People's Republic of China. Email: lxzhang@mail.hz.zj.cn

[2] Department of Statistics, Tunghai University, Taichung 40704, Taiwan

A two-parameter Wiener process is not of this type.

It is easy to see that $\mathsf{E}Z^2(R) = |R|^{2\alpha}$ for any $R = [x_1, x_2] \times [y_1, y_2]$, where $|\cdot|$ denotes the Lebesgue measure on $\mathbb{R}^2$, and

$$Z(R) = Z(x_2, y_2) - Z(x_2, y_1) - Z(x_1, y_2) + Z(x_1, y_1).$$

Let $0 < a_T \le T$, $b_T \ge \sqrt{T}$ be two functions of T. Define

$$
(1.1) \quad
\begin{cases}
L_T = \{R = [x_1, x_2] \times [y_1, y_2] : 0 \le x_1 \le x_2 \le b_T, \\
\qquad\quad 0 \le y_1 \le y_2 \le b_T, x_2 y_2 \le T, |R| \le a_T\}, \\
L_T^* = \{R : R \in L_T, |R| = a_T\}, \\
\delta_T = \{2a_T^{2\alpha}\big(\log(T/a_T) + \log(1 + \log(b_T/\sqrt{a_T})) + \log\log T\big)\}^{-1/2}.
\end{cases}
$$

In the book of Csörgő and Révész (1981), they showed "how big the increments of a two-parameter Wiener process $W(\cdot, \cdot)$ are?".

Theorem A. *Suppose $\alpha = 1/2$. Let $0 < a_T \le T$ and $b_T \ge \sqrt{T}$ are non-decreasing functions such that*

(i) *δ_T is non-decreasing,*
(ii) *T/a_T is non-decreasing.*

If, for any given $\epsilon > 0$, there exists $\theta_0 = \theta_0(\epsilon)$ such that

$$(1.2) \qquad \limsup_{k \to \infty} \delta_{\theta^k}/\delta_{\theta^{k+1}} \le 1 + \epsilon, \text{ for } 1 < \theta \le \theta_0,$$

then

$$(1.3) \qquad \limsup_{T \to \infty} \sup_{R \in L_T^*} \delta_T |W(R)| = \limsup_{T \to \infty} \sup_{R \in L_T} \delta_T |W(R)| = 1 \quad a.s.$$

Furthermore, if

$$(iii) \qquad \lim_{T \to \infty} \big\{ \log(T/a_T) + \log\big(1 + \log(b_T/\sqrt{a_T})\big) \big\}/\log\log T = \infty,$$

then

$$(1.4) \qquad \lim_{T \to \infty} \sup_{R \in L_T^*} \delta_T |W(R)| = \lim_{T \to \infty} \sup_{R \in L_T} \delta_T |W(R)| = 1 \quad a.s.$$

Lin (1984) and Zhang (1996b, 1997b) studied the liminfs when the condition (iii) is not satisfied. Almost all results of Csörgő and Révész (1981) on the increments of one-parameter Wiener process were extended to other and general one-parameter Gaussian processes (cf. Ortega 1984, Csáki, Csörgő and Shao 1992, Csörgő and Shao 1993, Monrad and Rootzen 1995, etc.). Also, some authors tried to extend Theorem A to other two-parameter Gaussian processes (cf. Kong 1989). But to the best of our knowledge, no pioneering results actually included Theorem A. Especially, the equalities in (1.3) were not obtained for other two-parameter Gaussian processes. Lately, Choi and Kôno (1999), Lin and Choi (1999) established some limit results on a kind of increments of a two-parameter fractional Lévy Brownian motion. But their results are far away from the type of Theorem A. The purpose of this paper is to extend Theorem A to a BESW process of order α. Our results read as follows.

Theorem 1.1. *Let* $0 < a_T \le T$ *and* $b_T \ge \sqrt{T}$ *be two functions of* T. *Suppose that* b_T *is quasi-increasing, i.e., for some* c_0, $b_{T_1} \le c_0 b_{T_2}$ *for all* $T_1 \le T_2$. *Then*

$$(1.5) \qquad \limsup_{T \to \infty} \sup_{R \in L_T^*} \delta_T Z(R) = \limsup_{T \to \infty} \sup_{R \in L_T} \delta_T |Z(R)| = 1 \quad a.s.$$

Furthremore, if the condition (iii) in Theorem A is satisfied, then

$$(1.6) \qquad \lim_{T \to \infty} \sup_{R \in L_T^*} \delta_T Z(R) = \lim_{T \to \infty} \sup_{R \in L_T} \delta_T |Z(R)| = 1 \quad a.s.$$

If the condition (iii) is not satisfied, we have the following result on liminfs:

Theorem 1.2. *Let* $0 < a_T \le T$ *and* $b_T \ge \sqrt{T}$ *be two functions of* T. *Suppose that* b_T *is quasi-increasing. Let*

$$(1.7) \quad \gamma_T = \left\{ 2a_T^{2\alpha} \left(\log(T/a_T) + \log\left(1 + \log(b_T/\sqrt{a_T})\right) - \log\log\log T \right) \right\}^{-1/2}.$$

Assume that

$$(\mathrm{iii})' \qquad \liminf_{T \to \infty} \frac{\log(T/a_T) + \log(1 + \log(b_T/\sqrt{a_T}))}{\log\log\log T} = r > 1.$$

If $r = +\infty$ *or* $0 < \alpha \le 1/2$ *or* $\liminf_{T \to \infty} a_T/T > 0$, *then*

$$(1.8) \qquad \liminf_{T \to \infty} \sup_{R \in L_T^*} \gamma_T Z(R) = \liminf_{T \to \infty} \sup_{R \in L_T} \gamma_T |Z(R)| = 1 \quad a.s.$$

Remark. In our Theorems 2.1 and 2.2, the conditions (i), (ii) and (1.2) in Theorem A are not used. Also, the functions a_T and b_T may be not non-decreasing.

If we take $a_T = T$ and $b_T = \sqrt{T}$, by Theorem 1.1, we have the following law of iterated logarithm.

$$\limsup_{T\to\infty} \frac{Z(\sqrt{T},\sqrt{T})}{\sqrt{2T^{2\alpha}\log\log T}} = \limsup_{T\to\infty} \sup_{0\le x,y\le \sqrt{T}} \frac{|Z(x,y)|}{\sqrt{2T^{2\alpha}\log\log T}} = 1 \quad \text{a.s.,}$$

i.e.,

$$\limsup_{T\to\infty} \frac{Z(T,T)}{\sqrt{2T^{4\alpha}\log\log T}} = \limsup_{T\to\infty} \sup_{0\le x,y\le T} \frac{|Z(x,y)|}{\sqrt{2T^{4\alpha}\log\log T}} = 1 \quad \text{a.s.}$$

Take $b_T = T$ and $a_T = T$. From Theorems 1.1 and 1.2, we have the following:

Corollary 1.1. *We have*

$$\limsup_{T\to\infty} \sup_{\substack{0\le x,y\le T \\ xy=T}} \frac{Z(x,y)}{\sqrt{4T^{2\alpha}\log\log T}} = \limsup_{T\to\infty} \sup_{\substack{0\le x,y\le T \\ xy\le T}} \frac{|Z(x,y)|}{\sqrt{4T^{2\alpha}\log\log T}} = 1 \quad \text{a.s.,}$$

$$\liminf_{T\to\infty} \sup_{\substack{0\le x,y\le T \\ xy=T}} \frac{Z(x,y)}{\sqrt{2T^{2\alpha}\log\log T}} = \liminf_{T\to\infty} \sup_{\substack{0\le x,y\le T \\ xy\le T}} \frac{|Z(x,y)|}{\sqrt{2T^{2\alpha}\log\log T}} = 1 \quad \text{a.s.}$$

If we take $b_T = T^{1/2}$ and write a_T to be $a_{1,T}a_{2,T}$, from (1.5) we can get that

$$\limsup_{T\to\infty} \frac{\Lambda(T^{1/2}; a_{1,T}, a_{1,T})}{\left\{2(a_{1,T}a_{2,T})^{2\alpha}\left(\log\frac{T}{a_{1,T}a_{2,T}} + \log\log T\right)\right\}^{1/2}} \le 1 \quad \text{a.s.,}$$

where $0 < a_{1,T} \le T^{1/2}$, $0 < a_{2,T} \le T^{1/2}$ and

$$(1.9) \quad \Lambda(U; a, b) = \sup_{0\le u\le U-a} \sup_{0\le v\le U-b} \sup_{0\le t\le a} \sup_{0\le s\le b} \left|W\left([u, u+t]\times[v, v+s]\right)\right|$$

for $0 < a, b \le U$. This is equivalent to the following corollary, which is corresponding to Theorem 1 of Choi and Kôno (1999) and Theorem 1 of Lin and Choi (1999).

Corollary 1.2. *Suppose that $0 < a_T \le T$ and $0 < b_T \le T$ are two functions of T. Let $\Lambda(U; a, b)$ be defined as in (1.9). Then we have*

$$\limsup_{T\to\infty} \frac{\Lambda(T; a_T, b_T)}{\left\{2(a_T b_T)^{2\alpha}\left(\log\frac{T}{a_T} + \log\frac{T}{b_T} + \log\log T\right)\right\}^{1/2}} \le 1 \quad \text{a.s.}$$

The proofs of Theorems 1.1 and 1.2 are based on the following inequalities and will be given in Section 3 and 4, respectively.

Theorem 1.3. *Let* $L_T^{**} = \{R = [x_1, x_2] \times [y_1, y_2] : R \in L_T^*, 0 \leq x_1 \leq x_2 \leq \sqrt{T}\}$. *Assume that* $0 < a_T \leq T$, $b_T \geq \sqrt{T}$. *Then, for any* $\delta > 0$, *there exists* $C = C(\delta) > 0$ *such that*

$$\mathsf{P}\Big\{ \sup_{R \in L_T^*} Z(R) \leq u a_T^\alpha \Big\} \leq \mathsf{P}\Big\{ \sup_{R \in L_T^{**}} Z(R) \leq u a_T^\alpha \Big\}$$

(1.10)
$$\leq \exp\Big\{ -C\Big(1 \vee \Big(\frac{T a_T^{-1} \log(b_T/\sqrt{T})}{k}\Big)\Big)e^{-u^2/(2-\delta)}\Big\}$$
$$+ \exp\{-Ck^{2-2\alpha}u^2\}$$

for any k *large enough. Furthermore, if* $0 < \alpha \leq 1/2$ *and* $a_T/T \leq \rho$ *for some* $0 < \rho < 1$, *then*

(1.11)
$$\mathsf{P}\Big\{ \sup_{R \in L_T^*} Z(R) \leq u a_T^\alpha \Big\} \leq \mathsf{P}\Big\{ \sup_{R \in L_T^{**}} Z(R) \leq u a_T^\alpha \Big\}$$
$$\leq \exp\big\{ -C T a_T^{-1} \log(b_T/\sqrt{T}) e^{-u^2/(2-\delta)}\big\},$$

and , if $a_T \geq \rho$ *for some* $0 < \rho \leq 1$, *then*

$$\mathsf{P}\Big\{ \sup_{R \in L_T^*} Z(R) \leq u a_T^\alpha \Big\} \leq \mathsf{P}\Big\{ \sup_{R \in L_T^{**}} Z(R) \leq u a_T^\alpha \Big\}$$

(1.12)
$$\leq \exp\Big\{ -C\Big(1 \vee \Big(\frac{T a_T^{-1} \log(b_T/\sqrt{T})}{k}\Big)\Big)e^{-u^2/(2-\delta)}\Big\}$$
$$+ \exp\big\{ -C e^{(2-2\alpha)k} u^2\big\}$$

for any k *large enough, where* $C = C(\delta, \rho)$ *depends on* δ *and* ρ *only.*

Theorem 1.4. *Let* $\{X(t, s); t, s \geq 0\}$ *be a mean zero continuous two-parameter Gaussian process with* $X(0, 0) = 0$ *a.s. and* $\mathsf{E}X^2(R) \leq \sigma^2(|R|)$, *where* $\sigma(x)$ *is a non-decreasing function such that* $\sigma(x)/x^\alpha$ *is quasi-increasing for some* $\alpha > 0$, *i.e., for some constant* c_0, $\sigma(x)/x^\alpha \leq c_0 \sigma(y)/y^\alpha$ *for all* $0 < x \leq y$. *Then for any given* $\epsilon > 0$, *there exists* $C = C(\epsilon)$ *such that*

$$\mathsf{P}\Big\{ \sup_{R \in L_T} |X(R)| \leq u \sigma(a_T) \Big\}$$

(1.13)
$$\geq \exp\Big\{ -C \frac{T}{a_T}\Big(1 + \log \frac{T}{a_T}\Big)\Big(1 + \log \frac{b_T}{\sqrt{a_T}}\Big)e^{-u^2/(2+\epsilon)}\Big\},$$

for all $u \geq 1$, $0 < a_T \leq T$ *and* $b_T \geq \sqrt{T}$.

Remark. Obviously, (1.13) implies

$$
(1.14) \qquad
\begin{aligned}
&\mathsf{P}\left\{ \sup_{R \in L_T} |X(R)| \ge u\sigma(a_T) \right\} \\
&\qquad \le C\frac{T}{a_T}\left(1 + \log \frac{T}{a_T}\right)\left(1 + \log \frac{b_T}{\sqrt{a_T}}\right) e^{-u^2/(2+\epsilon)}.
\end{aligned}
$$

This is an inequality corrosponding to (1.12.8) in Csörgő and Révész (1981). Also, if L_T is defined by

$$
\begin{aligned}
L_T = \{ R = [x_1, x_2] \times [y_1, y_2] : 0 \le x_1 \le x_2 \le b_T^{(1)}, \\
0 \le y_1 \le y_2 \le b_T^{(2)}, x_2 y_2 \le T, |R| \le a_T \},
\end{aligned}
$$

where $b_T^{(1)} b_T^{(2)} \ge T$, then (1.13) and (1.14) are also ture with $\sqrt{b_T^{(1)} b_T^{(2)}/a_T}$ taking the place of $b_T/\sqrt{a_T}$.

2. Proofs of Theorems 1.3 and 1.4.

To prove Theorem 1.3, we need the following lemma:

Lemma 2.1. *Let $\{Y(t); t \ge 0\}$ be a mean zero Gaussian process with $\mathsf{E}Y^2(t) = 1$ and $\mathsf{E}Y(t+k)Y(t) \le \rho(k)$, where $\rho(k) \ge 0$ is non-increasing and $\rho(k) \to 0$ $(k \to \infty)$. Then, for any $\epsilon > 0$, there exists $C = C(\epsilon)$ such that*

$$
\mathsf{P}\left\{ \sup_{0 \le t \le b} Y(t) \le u \right\} \le \exp\left\{ -C\left(1 \vee \frac{b}{k}\right) e^{-u^2/(2-\epsilon)} \right\} + \exp\left\{ -C\frac{u^2}{\rho(k)} \right\}
$$

for all $u \ge 1$ and k large enough.

Proof. Let $\xi_i = Y(ik)$ for $i = 0, \cdots [b/k]$. Then we have

$$
\mathsf{E}\xi_i \xi_j \le \rho(|i - j|k) \le \rho(k), \quad i \ne j.
$$

Let τ and $\eta_i, i = 0, \cdots, [b/k]$, be independent mean zero Gaussian random variables with $\mathsf{E}\tau^2 = \rho(k)$, $\mathsf{E}\eta_i^2 = 1 - \rho(k)$. Then $\mathsf{E}(\tau + \eta_i)^2 = 1 = \mathsf{E}\xi_i^2$ and

$$
\mathsf{E}\xi_i \xi_j \le \rho(k) = \mathsf{E}(\tau + \eta_i)(\tau + \eta_j), \quad i \ne j.
$$

By the Slepian lemma, we have

$$\mathsf{P}\Big\{\sup_{0\le t\le b} Y(t) \le u\Big\} \le \mathsf{P}\Big\{\max_{0\le i\le [b/k]} \xi_i \le u\Big\} \le \mathsf{P}\Big\{\max_{0\le i\le [b/k]} (\tau + \eta_i) \le u\Big\}$$

$$\le \mathsf{P}\Big\{\max_{0\le i\le [b/k]} \eta_i \le (1+\delta)u\Big\} + \mathsf{P}(\tau \ge \delta u)$$

$$= \Big\{1 - \mathsf{P}\Big(N(0,1) > \frac{(1+\delta)u}{\sqrt{1-\rho(k)}}\Big)\Big\}^{1+[b/k]} + \mathsf{P}\{N(0,1) \ge \delta u/\sqrt{\rho(k)}\}$$

$$\le \exp\Big\{-C(1+[b/k])\exp\Big\{-\frac{(1+\delta)u^2}{2(1-\rho(k))}\Big\}\Big\} + \exp\Big\{-\frac{\delta^2 u^2}{2\rho(k)}\Big\}$$

$$\le \exp\Big\{-C\Big(1 \vee \frac{b}{k}\Big)\exp\Big\{-\frac{u^2}{2-\epsilon}\Big\}\Big\} + \exp\Big\{-\frac{Cu^2}{\rho(k)}\Big\}$$

for δ small and k large enough.

Proof of Theorem 1.3. Let

$$U(t) = Z(\sqrt{T}e^{-t}, \sqrt{T}e^{t}) - Z((\sqrt{T} - a_T/\sqrt{T})e^{-t}, \sqrt{T}e^{t}).$$

Then we have

$$\mathsf{E}U(t)U(s) = \frac{1}{4}T^{2\alpha}e^{2\alpha(t-s)}\{1 + e^{-2\alpha(t-s)} - |1 - e^{-(t-s)}|^{2\alpha}\}$$

$$(2.1) \qquad \times \Big\{\Big|1 - \Big(1 - \frac{a_T}{T}\Big)e^{-(t-s)}\Big|^{2\alpha} + \Big|\Big(1 - \frac{a_T}{T}\Big) - e^{-(t-s)}\Big|^{2\alpha}$$

$$- |1 - e^{-(t-s)}|^{2\alpha} - \Big|\Big(1 - \frac{a_T}{T}\Big) - \Big(1 - \frac{a_T}{T}\Big)e^{-(t-s)}\Big|^{2\alpha}\Big\}.$$

If $a_T/T < 1$, let $d = -\log(1 - a_T/T)$ and $x = e^{-d} = 1 - a_T/T$. Then $0 < x < 1$
and

$$\mathsf{E}U(s)U(s+kd)$$

$$= \frac{1}{4}T^{2\alpha}e^{2\alpha kd}\{1 + e^{-2\alpha kd} - |1 - e^{-kd)}|^{2\alpha}\}$$

$$(2.2) \qquad \times \{(1 - e^{-(k+1)d})^{2\alpha} + (e^{-d} - e^{-kd})^{2\alpha}$$

$$- (1 - e^{-kd})^{2\alpha} - (e^{-d} - e^{-(k+1)d})^{2\alpha}\}$$

$$= \frac{1}{4}T^{2\alpha}x^{-2\alpha k}\{1 + x^{2\alpha k} - (1 - x^k)^{2\alpha}\}g(x),$$

where

$$g(x) = (1 - x^{k+1})^{2\alpha} + (x - x^k)^{2\alpha} - (1 - x^k)^{2\alpha} - (x - x^{k+1})^{2\alpha}.$$

Obviously, $1 + x^{2\alpha k} - (1 - x^k)^{2\alpha} > 0$. Let $f(t) = (1 + t)^{2\alpha} - t^{2\alpha}$. Then $f'(t) = 2\alpha\{(1+t)^{2\alpha-1} - t^{2\alpha-1}\}$. It follows that $|f'(t)| \le K_\alpha t^{2\alpha-2}$ $(t > 0)$, and $f'(t) \le 0$ $(t > 0)$ if $0 < \alpha \le 1/2$. Let $y = x + \cdots + x^{k-1} = (x - x^k)/(1 - x)$. Then

$$\begin{aligned}
g(x) &= (1 - x)^{2\alpha}\{(1 + x + \cdots + x^k)^{2\alpha} + (x + \cdots + x^{k-1})^{2\alpha} \\
&\quad - (1 + x + \cdots + x^{k-1})^{2\alpha} - (x + \cdots + x^k)^{2\alpha}\} \\
&= (1 - x)^{2\alpha}\{(1 + x + xy)^{2\alpha} + y^{2\alpha} - (1 + y)^{2\alpha} - (x + xy)^{2\alpha}\} \\
&= (1 - x)^{2\alpha}\{f(x + xy) - f(y)\} \\
&= (1 - x)^{2\alpha}(x + xy - y)f'(\xi) = (1 - x)^{2\alpha}x^k f'(\xi),
\end{aligned}$$

where $(x - x^k)/(1 - x) = y \le \xi \le x + xy = (x - x^{k+1})(1 - x)$. It follows that $g(x) \le 0$ if $0 \le \alpha \le 1/2$, and

$$|g(x)| \le (1 - x)^{2\alpha}x^k K_\alpha \xi^{2\alpha-2} \le (1 - x)^{2\alpha}x^k K_\alpha \left(\frac{1 - x}{x - x^k}\right)^{2-2\alpha}.$$

Obviously, $x^{-2\alpha k}\{1 + x^{2\alpha k} - (1 - x^k)^{2\alpha}\} \ge 0$ and $x^{-2\alpha k}\{1 + x^{2\alpha k} - (1 - x^k)^{2\alpha}\} \le K_\alpha x^{(1-2\alpha)k}$ when $1/2 \le \alpha < 1$. It follows that

$$EU(s)U(s + kd) \le 0, \quad 0 < \alpha \le 1/2,$$

and

$$\begin{aligned}
EU(s)U(s + kd) &\le K_\alpha T^{2\alpha}(1 - x)^{2\alpha}\left(\frac{(1 - x)x^k}{x - x^k}\right)^{2-2\alpha} \\
&= K_\alpha a_T^{2\alpha}\left(\frac{1}{x^{-1} + \cdots + x^{-(k-1)}}\right)^{2-2\alpha} \\
&\le K_\alpha a_T^{2\alpha} k^{-(2-2\alpha)}, \quad 1/2 \le \alpha < 1.
\end{aligned}$$

If $a_T/T = 1$, let $d = 1$. Then

$$\begin{aligned}
EU(s)U(s + kd) &= \frac{1}{4}T^{2\alpha}e^{2\alpha kd}\{1 + e^{-2\alpha kd} - (1 - e^{-kd})^{2\alpha}\}^2 \\
&\le K_\alpha T^{2\alpha}e^{-(2-2\alpha)kd} \le K_\alpha T^{2\alpha} k^{-(2-2\alpha)}.
\end{aligned}$$

Now, let $Y(t) = a_T^{-\alpha} U(td)$, where $d = 1$ if $a_T/T = 1$, $d = -\log(1 - a_T/T)$ if $a_T/T < 1$. Then

$$\mathsf{E}Y(t)Y(t+k) \leq \begin{cases} 0 & \text{if } 0 < \alpha \leq 1/2, \\ K_\alpha e^{-(2-2\alpha)kd} & \text{if } \frac{a_T}{T} = 1, \\ K_\alpha k^{-(2-2\alpha)} & \text{if } 1/2 \leq \alpha < 1 \end{cases}$$
$$=: \rho(k) \leq K_\alpha k^{-(2-2\alpha)}.$$

Case I. $a_T/T \leq \rho < 1$. Note that $-\log(1-t) \geq C_\rho t$ for $0 \leq t \leq \rho$. By Lemma 2.1, we have

$$\mathsf{P}\left(\sup_{R \in L_T^{**}} Z(R) \leq u a_T^\alpha \right)$$

$$\leq \mathsf{P}\left(\sup_{0 \leq t \leq \log(b_T/\sqrt{T})} U(t) \leq u a_T^\alpha \right) = \mathsf{P}\left(\sup_{0 \leq t \leq d^{-1} \log b_T/\sqrt{T}} Y(t) \leq u \right)$$

$$\leq \exp\left\{ -C\left(1 \vee \left(\frac{d^{-1}\log(b_T/\sqrt{T})}{k} \right) \right) e^{-u^2/(2-\epsilon)} \right\} + \exp\left\{ -\frac{Cu^2}{\rho(k)} \right\}$$

$$\leq \exp\left\{ -C\left(1 \vee \left(\frac{Ta_T^{-1}\log(b_T/\sqrt{T})}{k} \right) \right) e^{-u^2/(2-\epsilon)} \right\} + \exp\left\{ -Ck^{2-2\alpha}u^2 \right\}.$$

If $0 < \alpha \leq 1/2$, then

$$\mathsf{P}\left(\sup_{R \in L_T^{**}} Z(R) \leq u a_T^\alpha \right) \leq \mathsf{P}\left(\sup_{0 \leq t \leq d^{-1} \log b_T/\sqrt{T}} Y(t) \leq u \right)$$

$$\leq \exp\left\{ -CTa_T^{-1}\log(b_T/\sqrt{T})e^{-u^2/2} \right\}.$$

Case II. $a_T/T \geq \rho$. First we assume that $a_T/T = 1$. Then

$$\mathsf{P}\left(\sup_{R \in L_T^{**}} Z(R) \leq u a_T^{2\alpha} \right)$$

$$\leq \exp\left\{ -C\left(1 \vee \left(\frac{d^{-1}\log(b_T/\sqrt{T})}{k} \right) \right) e^{-u^2/(2-\epsilon)} \right\} + \exp\left\{ -\frac{Cu^2}{\rho(k)} \right\}$$

$$\leq \exp\left\{ -C\left(1 \vee \left(\frac{\log(b_T/\sqrt{T})}{k} \right) \right) e^{-u^2/(2-\epsilon)} \right\} + \exp\{-Ce^{-(2-2\alpha)k}u^2\}.$$

Now assume that $\rho \le a_T/T \le 1$. Then

$$P\left(\sup_{R \in L_T^{**}} Z(R) \le u a_T^\alpha\right) \le P\left(\sup_{\substack{xy \le a_T, 0 \le x \le \sqrt{a_T}, \\ 0 \le x, y \le b_T}} Z(R) \le u a_T^\alpha\right)$$

$$\le \exp\left\{-C\left(1 \vee \left(\frac{\log(b_T/\sqrt{a_T})}{k}\right)\right)e^{-u^2/(2-\epsilon)}\right\} + \exp\{-Ce^{-(2-2\alpha)k}u^2\}$$

$$\le \exp\left\{-C\left(1 \vee \left(\frac{T a_T^{-1}\log(b_T/\sqrt{T})}{k}\right)\right)e^{-u^2/(2-\epsilon)}\right\} + \exp\{-Ce^{-(2-2\alpha)k}u^2\}.$$

Theorem 1.3 is proved. This completes the proof.

Remark. If one can show that $EU(t)U(s)$ in (2.1) is a non-increasing convex function of $t - s$ or $EU(s)U(s + kd)$ in (2.2) is a non-increasing convex function of k, then, similarly to the proof of Lemma 1 of Zhang (1997a), one can show that

$$P\left(\sup_{R \in L_T^{**}} |Z(R)| \le u a_T^\alpha\right) \le \exp\left\{-C\left(1 \vee \left(\frac{T a_T^{-1}\log(b_T/\sqrt{T})}{k}\right)\right)e^{-u^2/(2-\epsilon)}\right\},$$

for k large enough.

To prove Theorem 1.4, we need some lemmas:

Lemma 2.2. (KHATRI 1967, ŠIDÁK 1968) *Let* $\{Y_i; i = 1, \cdots, N\}$ *be a mean zero Gaussian vector in* $\mathbb{R}^N$. *Then*

$$P\left\{\max_{i \le N} |Y_i| \le \lambda_i\right\} \ge \prod_i^N P(|Y_i| \le \lambda_i),$$

for all $\lambda_i \ge 0$, $i = 1, \cdots, N$.

Let $\{X(t); t \in T\}$ be a mean zero Gaussian process and $x(t) > 0$ be a real function. If there are a mean zero Gaussian process $U(t)$ on a countable set T_c and a function $u(t) > 0$ $(t \in T_c)$ such that

$$\left\{\sup_{t \in T} \frac{|Y(t)|}{x(t)} \le 1\right\} \supset \left\{\sup_{t \in T_c} \frac{|U(t)|}{u(t)} \le 1\right\} \quad \text{a.s.,}$$

then, by Lemma 2.2, we have

$$P\left\{\sup_{t \in T} \frac{|Y(t)|}{x(t)} \le 1\right\} \ge \prod_{t \in T_c} P\left\{\frac{|U(t)|}{u(t)} \le 1\right\} =: P_X.$$

If such $U(\cdot)$ and $u(\cdot)$ exist and $U(\cdot)$ also satisfies that for any $t \in T_c$, $U(t)$ is a linear combination of $Y(s_1), \cdots, Y(s_m)$ for some $s_1, \cdots, s_m \in T$, then we call P_X a KS *lower bound* (KSLB) of $P(\sup_{t \in T} \frac{|X(t)|}{x(t)} \le 1)$ and write

$$P\left\{ \sup_{t \in T} \frac{|X(t)|}{x(t)} \le 1 \right\} \overset{\text{KS}}{\ge} P_X.$$

Obviously, by Lemma 2.2, we have

Lemma 2.3. *Let* T_i, $i = 1, 2, \cdots$ *be parameter sets,* $\{Y_i(t), t \in T_i, i = 1, 2, \cdots\}$ *be a Gaussian process with mean zero. Assume that*

$$P\left(\sup_{t \in T_i} \frac{|Y_i(t)|}{x_i(t)} \le 1 \right) \overset{\text{KS}}{\ge} p_i \quad i = 1, 2, \cdots.$$

Then

$$P\left(\sup_i \sup_{t \in T_i} \frac{|Y_i(t)|}{x_i(t)} \le 1 \right) \overset{\text{KS}}{\ge} \prod_{i=1}^{\infty} p_i,$$

where $x_i(t) > 0$ $(t \in T_i)$ $(i = 1, 2, \cdots)$.

Lemma 2.4. (ZHANG 1996A) *Let* $\{\Gamma(t), -\infty < t < \infty\}$ *be an almost surely continuous Gaussian process with mean zero. Assume that there exists a nondereasing function* $u(h)$ *on* $[0, \infty)$ *such that*

$$E(\Gamma(t + h) - \Gamma(t))^2 \le u^2(h) \quad \text{for all } t \ge 0, h \ge 0.$$

And aussume that $u(x)/x^\alpha$ *is quasi-increasing for some* $\alpha > 0$. *Then there exists a constant* $C_\alpha > 0$ *such that*

$$P\left(\sup_{0 \le t \le T} \sup_{0 \le s \le h} |\Gamma(t + s) - \Gamma(t)| \le xu(h) \right)$$
$$\overset{\text{KS}}{\ge} \exp\left\{ -C_\alpha \left(\frac{T}{h} + 1 \right) x^{4/\alpha - 1} e^{-x^2/2} \right\}$$

for all $x \ge 1$ *and* $0 < h \le T < \infty$.

Repeating the proof of Lemma 2.4 (cf. Zhang 1996a), one can prove a similar inequality for a two-parameter Gaussian process (cf. the proof of Lemma 2.4 of Zhang 1997b).

Lemma 2.5. *Let $\{X(t,s); t,s \geq 0\}$ be a mean zero continuous two-parameter Gaussian process with $X(0,0) = 0$ a.s. and $EX^2(R) \leq \sigma^2(|R|)$, where $\sigma(x)$ is a non-decreasing function such that $\sigma(x)/x^\alpha$ is quasi-increasing for some $\alpha > 0$. Then, for any given $\epsilon > 0$, there exists $C = C(\epsilon)$ such that*

$$P\left\{ \sup_{\substack{x_0 \leq x \leq T_1 + x_0 \ 0 \leq s \leq h_1 \\ y_0 \leq y \leq T_2 + y_0 \ 0 \leq t \leq h_2}} |X([x, x+s] \times [y, y+t])| \leq \sigma(h_1 h_2) \right\}$$
$$\overset{\mathrm{KS}}{\geq} \exp\left\{ -C\left(\frac{T_1}{h_1} + 1\right)\left(\frac{T_2}{h_2} + 1\right) e^{-u^2/(2+\epsilon)} \right\}$$

holds for all $x_0, y_0 \geq 0$, $0 < h_1 \leq T_1$, $0 < h_2 \leq T_2$ and $u \geq 1$. Particularly,

$$P\left\{ \sup_{R \subset [x_0, x_0+T_1] \times [y_0, y_0+T_2]} |W(R)| \leq u(T_1 T_2)^{1/2} \right\} \overset{\mathrm{KS}}{\geq} \exp\{-ce^{-u^2/(2+\epsilon)}\}$$

for all $u \geq 1$, $T_1, T_2 > 0$ and $x_0, y_0 \geq 0$, where $R = [x_1, x_2] \times [y_1, y_2]$.

Proof of Theorem 1.4.. For any positive integer q, let $Q = Q(q) = 2^q$. We shall use the notations and concepts of Theorem 1.12.6 and its proofs in Csörgő and Révész (1981) such as $R_i(q)$, $R_i(j,l)$, $L_T^*(q)$ and $\tilde{L}_T^*(q)$, etc. First we have

$$\sup_{R \in L_T} |X(R)| \leq \sup_{R \in L_T^*(q)} \sup_{S \subset R} |X(S)| + 4 \sum_{i=0}^{\infty} \sup_{R \in \tilde{L}_T^*(q+i)} \sup_{S \subset R} |X(S)|,$$

where S is a rectangle with edges parallel to the coordinate axes. By Lemmas 2.3 and 2.5 , for any $\delta > 0$, there exists a constant $C_\delta > 0$ such that

$$P\left(\sup_{R \in L_T^*(q)} \sup_{S \subset R} |X(S)| \leq x\sigma(a_T) \right) \overset{\mathrm{KS}}{\geq} \exp\{-C_\delta \operatorname{Card} L_T^*(q) e^{-x^2/(2+\delta)}\},$$

$$P\left(\sup_{R \in \tilde{L}_T^*(q+i)} \sup_{S \subset R} |X(S)| \leq y_i \sigma(6a_T Q^{-1} 2^{-i}) \right)$$
$$\overset{\mathrm{KS}}{\geq} \exp\{-C_\delta \operatorname{Card} \tilde{L}_T^*(q+i) e^{-y_i^2/(2+\delta)}\}$$

for $x \geq 1$, $y_i \geq 1$. Noting that $\operatorname{Card} \tilde{L}_T^*(q+i) \leq 4 \operatorname{Card} L_T^*(q+i)$ and using Lemma 2.3 again, we have

$$P\left(\sup_{R \in L_T} |X(R)| \leq x\sigma(a_T) + 4\sum_{i=0}^{\infty} y_i \sigma(6a_T Q^{-1} 2^{-i}) \right)$$
$$\overset{\mathrm{KS}}{\geq} \exp\left\{ -C_\delta\left(\operatorname{Card} L_T^*(q) e^{-x^2/(2+\delta)} + 4\sum_{i=0}^{\infty} \operatorname{Card} L_T^*(q+i) e^{-y_i^2/(2+\delta)} \right) \right\}.$$

Let $y_i = (3i(2+\delta) + x^2)^{1/2}$ $(i = 0, 1, \cdots)$, we have

$$x\sigma(a_T) + 4\sum_{i=0}^{\infty} y_i\sigma(6a_TQ^{-1}2^{-i})$$

$$\leq x\sigma(a_T) + \sum_{i=0}^{\infty}(\sqrt{9i} + x)c_0(6Q^{-1}2^{-i})^{\alpha}\sigma(a_T)$$

$$\leq x\sigma(a_T)\left(1 + \sum_{i=0}^{\infty} c_0(6Q^{-1})^{\alpha}2^{-\alpha i}\right) + \sigma(a_T)c_0(6Q^{-1})^{\alpha}\sum_{i=0}^{\infty}\sqrt{9i}2^{-i\alpha}$$

$$= x\sigma(a_T)(1 + AQ^{-\alpha}) + \sigma(a_T)BQ^{-\alpha} \leq x\sigma(a_T)(1 + \epsilon)$$

for Q large enough and $x \geq 1$. Also, by Lemma 1.12.1 of Csörgő and Révész (1981) we have

$$Card\, L_T^*(q)e^{-x^2/(2+\delta)} + 4\sum_{i=0}^{\infty} Card\, L_T^*(q+i)e^{-y_i^2/(2+\delta)}$$

$$\leq e^{-x^2/(2+\delta)}\left(8Q^3 + 4\cdot 8Q^3\sum_{i=0}^{\infty}(2/e)^{3i}\right)\frac{T}{a_T}\left(1 + \log\frac{T}{a_T}\right)\left(1 + \log\frac{b_T}{\sqrt{a_T}}\right)$$

$$\leq CQ^3\frac{T}{a_T}\left(1 + \log\frac{T}{a_T}\right)\left(1 + \log\frac{b_T}{\sqrt{a_T}}\right)e^{-x^2/(2+\delta)}.$$

It follows that

$$\mathsf{P}\left\{\sup_{R\in L_T} |X(R)| \leq (1 + \epsilon)x\sigma(a_T)\right\}$$

$$\geq \exp\left\{-C\frac{T}{a_T}\left(1 + \log\frac{T}{a_T}\right)\left(1 + \log\frac{b_T}{\sqrt{a_T}}e^{-x^2/(2+\delta)}\right)\right\},$$

which completes the proof of Theorem 1.4.

3. Proof of Theorem 1.1

The proof will be completed by three steps.

Step 1. *We have*

$$(3.1) \qquad\qquad \limsup_{T\to\infty}\delta_T\sup_{R\in L_T} |Z(R)| \leq 1 \quad a.s.$$

180   ZHANG L. X., LU C. R. AND WANG Y. H.

Proof. Let

$$C_T = \frac{T}{a_T}\left(1 + \log\frac{T}{a_T}\right)\left(1 + \frac{b_T}{\sqrt{a_T}}\right),$$

$$\overline{\delta}_T = \left\{2a_T^{2\alpha}\left(\log C_T + \log\log(a_T^{\alpha} + a_T^{-\alpha})\right)\right\}^{-1/2}.$$

Obviously, $\limsup_{T\to\infty} \overline{\delta}_T^{-1}\delta_T \le 1$. So it is enough to show that

$$(3.2) \qquad\qquad \limsup_{T\to\infty} \overline{\delta}_T \sup_{R\in L_T} |Z(R)| \le 1 \quad \text{a.s.}$$

In the sequel of this paper, we write $a(T) = a_T$, $b(T) = b_T$, $C(T) = C_T$, etc. Let $1 < \theta < 2$. Define

$$(3.3) \qquad A_k = \left\{\begin{array}{ll} \{T : \theta^k \le a_T^{\alpha} < \theta^{k+1}\}, & -\infty < k < \infty, \\ B_i = \{T : 2^i \le C_T < 2^{i+1}\}, & i \ge 0, \\ T_{k,i} = \sup\{T : T \in A_k B_i\}, & \\ T_{k,i}^* = \inf\{T : T \in A_k B_i\}, & \\ b_{k,i} = \sup\{b(T) : T \in A_k B_i\}. & \end{array}\right.$$

It is easy to see that $b_{k,i} \ge \sqrt{T_{k,i}}$. Let

$$(3.4) \qquad \begin{aligned} L_{k,i} = \big\{R = [x_1, x_2] \times [y_1, y_2] : {}& 0 \le x_1 \le x_2 \le b_{k,i}, \\ & 0 \le y_1 \le y_2 \le b_{k,i}, x_2 y_2 \le T_{k,i}, |R| \le \theta^{k+1}\big\}. \end{aligned}$$

Then

$$(3.5) \qquad \begin{aligned} \limsup_{T\to\infty} \sup_{R\in L_T} \overline{\delta}_T |Z(R)| &\le \limsup_{|k|+l\to\infty} \sup_{i\ge l} \sup_{T\in A_{k,i}} \sup_{R\in L_T} \overline{\delta}_T |Z(R)| \\ &\le \limsup_{|k|+l\to\infty} \sup_{i\ge l} \sup_{R\in L_{k,i}} |Z(R)| / \left\{2\theta^{2k}\left(\log 2^i + \log\log\theta^{|k|}\right)\right\}^{1/2}. \end{aligned}$$

By the remark of Theorem 1.4, we have

$$(3.6) \qquad \begin{aligned} \mathsf{P}&\left(\sup_{R\in L_{k,i}} |Z(R)| / \left\{2\theta^{2k}\left(\log 2^i + \log\log\theta^{|k|}\right)\right\}^{1/2} \ge \theta^2\right) \\ &\le C\frac{T_{k,i}}{\theta^{k+1}}\left(1 + \log\frac{T_{k,i}}{\theta^{k+1}}\right)\left(1 + \log\frac{b_{k,i}}{\theta^{k+1}}\right) \\ &\qquad \times \exp\left\{-\frac{2\theta^4\theta^{2k}(\log 2^i + \log\log\theta^{|k|})}{(2+\epsilon)\theta^{2(k+1)}}\right\} \\ &\le C\frac{T_{k,i}}{\theta^{k+1}}\left(1 + \log\frac{T_{k,i}}{\theta^{k+1}}\right)\left(1 + \log\frac{b_{k,i}}{\theta^{k+1}}\right)\left(2^i\log\theta^{|k|}\right)^{-\theta} \\ &=: C_{k,i}\left(2^i\log\theta^{|k|}\right)^{-\theta}. \end{aligned}$$

If $T_{k,i} \in A_k B_i$, then $b_{k,i} \le c_o b(T_{k,i})$ since $b(T)$ is quasi-increasing. So

$$C_{k,i} \le C \frac{T_{k,i}}{a(T_{k,i})} \left(1 + \log \frac{T_{k,i}}{\sqrt{a(T_{k,i})}}\right)\left(1 + \log \frac{b_{k,i}}{\sqrt{a(T_{k,i})}}\right)$$
$$= CC(T_{k,i}) \le C2^{i+1}.$$

If $T_{k,i} \notin A_k B_i$, then exists a sequence $\{T_m\} \subset A_k B_i$ such that $T_m \to T_{k,i}$ and $b(T_m) \to b_{k,i}$. Then

$$C_{k,i} = C \lim_{m \to \infty} \frac{T_m}{\theta^{k+1}} \left(1 + \log \frac{T_m}{\theta^{k+1}}\right)\left(1 + \log \frac{b(T_m)}{\theta^{k+1}}\right)$$
$$\le \limsup_{m \to \infty} CC(T_m) \le C2^{i+1}.$$

It follows that

$$(3.7) \qquad\qquad C_{k,i} \le C2^i.$$

From (3.6), it follows that

$$
\sum_{|k|=1}^{\infty}\sum_{l=1}^{\infty} \mathsf{P}\left(\sup_{i \ge l}\ \sup_{R \in L_{k,i}} |Z(R)|/\left\{2\theta^{2k}\left(\log 2^i + \log\log\theta^{|k|}\right)\right\}^{1/2} \ge \theta^2\right)
$$
$$
\le \sum_{|k|=1}^{\infty}\sum_{l=1}^{\infty}\sum_{i=l}^{\infty} C2^i \left(2^i \log\theta^{|k|}\right)^{-\theta}
$$
$$
(3.8) \qquad \le C \sum_{|k|=1}^{\infty}\sum_{l=1}^{\infty}\sum_{i=l}^{\infty} 2^{(1-\theta)i}|k|^{-\theta}
$$
$$
\le C \sum_{l=1}^{\infty} 2^{-(\theta-1)l} \sum_{|k|=1}^{\infty} |k|^{-\theta} < \infty,
$$

which together with the Borel-Cantelli leamma implies

$$(3.9) \qquad \limsup_{|k|+l \to \infty}\ \sup_{i \ge l}\ \sup_{R \in L_{k,i}} |Z(R)|/\left\{2\theta^{2k}\left(\log 2^i + \log\log\theta^{|k|}\right)\right\}^{1/2} \le \theta^2.$$

By (3.5), (3.9) and the arbitrariness of $1 < \theta < 2$, (3.2) is proved.

Step 2. *We have*

$$(3.10) \qquad \limsup_{T \to \infty} \sup_{R \in L_T^*} \delta_T Z(R) \geq \limsup_{T \to \infty} \sup_{R \in L_T^{**}} \delta_T Z(R) \geq 1 \quad a.s.,$$

where L_T^{**} is defined as in Theorem 1.3.

To prove Step 2, we need one more lemma. First, it is easy to see that $\{Z(t,s);\ t,s \geq 0\}$ has the same distribution as

$$\left\{ \int_{-\infty}^{\infty} \int_{-\infty}^{\infty} \frac{1}{K_2^2} \left\{ |x-t|^{\frac{2\alpha-1}{2}} - |x|^{\frac{2\alpha-1}{2}} \right\} \left\{ |y-s|^{\frac{2\alpha-1}{2}} - |y|^{\frac{2\alpha-1}{2}} \right\} W(dx,dy) : t,s \geq 0 \right\},$$

where $\{W(x,y); -\infty < x,y < \infty\}$ is a standard two-parameter Wiener process and $K_\alpha^2 = \int_{-\infty}^{\infty} \left\{ |x-1|^{\frac{2\alpha-1}{2}} - |x|^{\frac{2\alpha-1}{2}} \right\}^2 dx$. So we can write

$$Z(t,s) = \int_{-\infty}^{\infty} \int_{-\infty}^{\infty} \frac{1}{K_\alpha^2} \left\{ |x-t|^{\frac{2\alpha-1}{2}} - |x|^{\frac{2\alpha-1}{2}} \right\} \left\{ |y-s|^{\frac{2\alpha-1}{2}} - |y|^{\frac{2\alpha-1}{2}} \right\} W(dx,dy).$$

Let $d_n = ne^{n^p}$, $A_n = (d_{n-1}/e^{n^p}, d_n/e^{n^p})$, where $p > 1$ and $p(1 - \epsilon) < 1$. Define

$$(3.11) \quad \begin{cases} X_n(t,s) = \displaystyle\int_{|x| \in (d_{n-1}, d_n)} \int_{-\infty}^{\infty} \frac{1}{K_\alpha^2} \left\{ |x-t|^{\frac{2\alpha-1}{2}} - |x|^{\frac{2\alpha-1}{2}} \right\} \\[2mm] \qquad\qquad \times \left\{ |y-s|^{\frac{2\alpha-1}{2}} - |y|^{\frac{2\alpha-1}{2}} \right\} W(dx,dy), \\[4mm] \overline{X}_n(t,s) = \displaystyle\int_{|x| \notin (d_{n-1}, d_n)} \int_{-\infty}^{\infty} \frac{1}{K_\alpha^2} \left\{ |x-t|^{\frac{2\alpha-1}{2}} - |x|^{\frac{2\alpha-1}{2}} \right\} \\[2mm] \qquad\qquad \times \left\{ |y-s|^{\frac{2\alpha-1}{2}} - |y|^{\frac{2\alpha-1}{2}} \right\} W(dx,dy), \\[4mm] \overline{Y}_n(t,s) = \displaystyle\int_{|x| \notin A_n} \int_{-\infty}^{\infty} \frac{1}{K_\alpha^2} \left\{ |x-t|^{\frac{2\alpha-1}{2}} - |x|^{\frac{2\alpha-1}{2}} \right\} \\[2mm] \qquad\qquad \times \left\{ |y-s|^{\frac{2\alpha-1}{2}} - |y|^{\frac{2\alpha-1}{2}} \right\} W(dx,dy), \end{cases}$$

Then $Z(t,s) = X_n(t,s) + \overline{X}_n(t,s)$, $\{\overline{X}_n(\cdot,\cdot)\}_{n=\infty}^{\infty}$ are independent, and

$$\{\overline{X}_n(t,s); t,s \geq 0\} \overset{\mathcal{D}}{=} \{ (e^{n^p})^\alpha \overline{Y}_n(t/e^{n^p}, s); t,s \geq 0 \}.$$

Lemma 3.1. *Let $\gamma > 0$ and $p > 1$. Then there exists a constant $c > 0$ depending on α, γ, p such that*

$$\mathsf{E}\overline{Y}_n^2\big([t, t + h_1] \times [s + h_2]\big) \leq c(h_1 h_2)^{2\alpha}\left(\log \frac{1}{h_1}\right)^{\gamma} n^{-\delta}$$

$$\leq c(h_1 h_2)^{2\alpha}\left(\log \frac{1}{h_1 h_2}\right)^{\gamma} n^{-\delta}$$

$$=: \overline{\sigma}(h_1 h_2)$$

uniformly in $n \geq 1$, $0 \leq t \leq n/2$, $0 \leq h_1, h_2 \leq 1$ and $s \geq 0$.

Proof. It follows that

$$\mathsf{E}\overline{Y}_n(t, s) = \int_{|x| \notin A_n} \int_{-\infty}^{\infty} \frac{1}{K_\alpha^4} \{|x - t|^{\frac{2\alpha-1}{2}} - |x|^{\frac{2\alpha-1}{2}}\}^2$$

$$\{|y - s|^{\frac{2\alpha-1}{2}} - |y|^{\frac{2\alpha-1}{2}}\}^2 \, dx dy$$

$$= h^{2\alpha} \int_{|x| \notin A_n} \frac{1}{K_\alpha^2} \{|x - t|^{\frac{2\alpha-1}{2}} - |x|^{\frac{2\alpha-1}{2}}\}^2 \, dx$$

$$\leq c(h_1 h_2)^{2\alpha}\left(\log \frac{1}{h_1}\right)^{\gamma} n^{-\delta}$$

by Lemma 2.9 of Zhang (1996a) and its proof.

Proof of Step 2. From Theorem 1.3, it follows that, for any $\epsilon > 0$, there exists $C = C(\epsilon)$ such that

$$\mathsf{P}\left(\sup_{R \in L_T^{**}} \delta_T Z(R) \leq 1 - 2\epsilon\right)$$

$$\leq \exp\left\{-C\frac{T}{a_T}\log\frac{b_T}{\sqrt{T}}\right.$$

$$\times \exp\left(-\frac{2(1 - 2\epsilon)^2(\log\frac{T}{a_T} + \log(1 + \log\frac{b_T}{\sqrt{T}}) + \log\log T)}{2 - \epsilon/2}\right)\right\}$$

$$\left. + \exp\left\{-4\left(\log\frac{T}{a_T} + \log\left(1 + \log\frac{b_T}{\sqrt{T}}\right) + \log\log T\right)\right\}\right\}$$

$$\leq \exp\left\{-C\left(\frac{T}{a_T}\log\frac{b_T}{\sqrt{T}}\right)^{\epsilon}(\log T)^{-(1-\epsilon)}\right\} + (\log T)^{-4}$$

(3.12)

for any $0 < a_T \leq T$ and $b_T \geq \sqrt{T}$.

We first assume that

$$\limsup_{T\to\infty} \frac{T a_T^{-1} \log(1 + \log(b_T/\sqrt{T}))}{\log\log T} = \infty.$$

Then there exists a sequence $T_n \uparrow \infty$ such that

$$\frac{T_n a^{-1}(T_n) \log(1 + \log b(T_n)/\sqrt{T_n})}{\log\log T_n} \to \infty.$$

It follows from (3.12) that, for n large enough,

$$\mathsf{P}\left(\sup_{R\in L_{T_n}^{**}} \delta_{T_n} Z(R) \leq 1 - 2\epsilon\right) \leq \exp\{-c(\log T_n)^2\} + (\log T_n)^{-4} \to 0, \quad (n \to \infty).$$

Thus we have

$$\limsup_{T\to\infty} \sup_{R\in L_T^{**}} \delta_T Z(R) \geq \limsup_{n\to\infty} \sup_{R\in L_{T_n}^{**}} \delta_{T_n} Z(R) \geq 1 - 2\epsilon \quad \text{a.s.}$$

which implies (3.10). Now we assume that

$$(3.13) \qquad \frac{T a_T^{-1} \log(1 + \log(b_T/\sqrt{T}))}{\log\log T} < r < \infty.$$

Let $T_n = e^{n^p}$, $p > 1$, $p(1 - \epsilon) < 1$. Let d_n, A_n, $X_n(\cdot,\cdot)$, $\overline{X}_n(\cdot,\cdot)$, $\overline{Y}_n(\cdot,\cdot)$ be difined as in (3.11). Then $d_n = nT_n$, $A_n = (d_{n-1}T_n^{-1}, d_n T_n^{-1})$ and

$$\{\overline{X}_n(t,s); t,s \geq 0\} \overset{\mathcal{D}}{=} \{T_n^\alpha \overline{Y}_n(tT_n^{-1}, s); t,s \geq 0\}.$$

Let

$$\overline{L}_{T_n}^{**} = \{R = [x_1, x_2] \times [y_1, y_2] : 0 \leq x_1 \leq x_2 \leq \sqrt{T_n}/T_n,$$
$$0 \leq y_1 \leq y_2 \leq b(T_n), x_2 y_2 \leq 1, |R| \leq a(T_n)/T_n\}.$$

By Lemma 3.1, $\mathsf{E}\overline{Y}_n^2(R) \leq \overline{\sigma}(|R|)$ for any $R \in \overline{L}_{T_n}^{**}$, where $\overline{\sigma}(t) = cn^{-\delta}t^{2\alpha}\left(\log\frac{1}{t}\right)^\gamma$.

From the remark of Theorem 1.4, it follows that

$$P\left(\sup_{R\in L_{T_n}^{**}} \delta(T_n)|\overline{X}_n(R)| \geq \epsilon\right) = P\left(\sup_{R\in \overline{L}_{T_n}^{**}} \delta(T_n)T_n^{\alpha}|\overline{Y}_n(R)| \geq \epsilon\right)$$

$$\leq C\frac{T_n}{a(T_n)}\left(1 + \log\frac{T_n}{a(T_n)}\right)\left(1 + \log\sqrt{\frac{b(T_n)\sqrt{T_n}/T_n}{a(T_n)/T_n}}\right)$$

$$\times \exp\left\{ - \frac{2\epsilon^2\delta^{-2}(T_n)T_n^{-2\alpha}}{(2+\epsilon)cn^{-\delta}\left(\frac{a(T_n)}{T_n}\right)^{2\alpha}\left(\log\frac{T_n}{a(T_n)}\right)^{\gamma}} \right\}$$

$$\leq C\frac{T_n}{a(T_n)}\left(1 + \log\frac{T_n}{a(T_n)}\right)\left(1 + \log\sqrt{\frac{b(T_n)\sqrt{T_n}/T_n}{a(T_n)/T_n}}\right)$$

$$\times \exp\left\{ - c\frac{\log\frac{T_n}{a(T_n)} + \log\left(1 + \log\frac{b(T_n)}{\sqrt{a(T_n)}}\right) + \log\log T_n}{n^{-\delta}(\log n)^{\gamma}} \right\}$$

$$\leq (\log T_n)^{-3} \leq n^{-2}$$

(3.14)

for n large enough. It follows from the Borel-Cantelli lemma that

$$\limsup_{n\to\infty} \sup_{R\in L_{T_n}^{**}} \delta(T_n)|\overline{X}_n(R)| \leq \epsilon \quad \text{a.s.} \tag{3.15}$$

On the other hand, by (3.12), we have

$$P\left\{ \sup_{R\in L_T^{**}} \delta_T Z(R) \geq 1 - 2\epsilon \right\}$$

$$\geq 1 - \exp\left\{ - C\left(\frac{T}{a_T}\log\frac{b_T}{\sqrt{T}}\right)^{\epsilon}(\log T)^{-(1-\epsilon)} \right\} - (\log T)^{-4}$$

$$\geq c(\log T)^{-(1-\epsilon)} - (\log T)^{-4}.$$

It follows that

$$\sum_{n=1}^{\infty} P\left\{ \sup_{R\in L_{T_n}^{**}} \delta(T_n)X_n(R) \geq 1 - 3\epsilon \right\}$$

$$\geq \sum_{n=1}^{\infty} \left(P\left\{ \sup_{R\in L_{T_n}^{**}} \delta(T_n)Z(R) \geq 1 - 2\epsilon \right\} - P\left\{ \sup_{R\in L_{T_n}^{**}} \delta(T_n)|\overline{X}_n(R)| \geq \epsilon \right\} \right)$$

$$\geq \sum_{n=1}^{\infty} \{cn^{-p(1-\epsilon)} - cn^{-4p} - n^{-2}\} = +\infty,$$

which together with the Borel-Cantelli lemma and the independence of $\{X_n(\cdot,\cdot)\}$ implies that

$$(3.16) \qquad \limsup_{n\to\infty} \sup_{R\in L^{**}_{T_n}} \delta(T_n)X_n(R) \geq 1-3\epsilon.$$

Putting (3.15) and (3.16) together yields

$$\limsup_{T\to\infty} \sup_{R\in L^{**}_{T}} \delta_T Z(R) \geq \limsup_{n\to\infty} \sup_{R\in L^{**}_{T_n}} \delta(T_n)Z(R) \geq 1-4\epsilon \quad \text{a.s.},$$

which implies (3.10). Combining (3.1) and (3.10) implies (1.5).

Step 3. *If the condition* (iii) *is satisfied, then*

$$(3.17) \qquad \liminf_{T\to\infty} \sup_{R\in L^{*}_{T}} \delta_T Z(R) \geq 1 \quad \text{a.s.}$$

Proof. By Theorem 1.2, (3.17) is ture.

4. Proof of Theorem 1.2

The proof of Theorem 1.2 will be given by two propositions.

Proposition 4.1. *Let* $0 < a_T \leq T$ *and* $b_T \geq \sqrt{T}$ *be two functions of* T *such that* b_T *is quasi-increasing. Let* $\Delta_T = Ta_T^{-1}(1+\log Ta_T^{-1})(1+\log(b_T a_T^{-1/2}))$. *If*

(iv) $\Delta_T/\log\log T \to \infty$ *as* $T\to\infty$,

then

$$(4.1) \qquad \liminf_{T\to\infty} \sup_{R\in L_T} \gamma_T^{(1)}|Z(R)| \leq 1 \quad \text{a.s.},$$

where $\gamma_T^{(1)} = \{2a_T^{2\alpha}(\log\Delta_T - \log\log\log T)\}^{-1/2}$.

Proof. By Theorem 1.4, we have

$$
\begin{aligned}
&\mathsf{P}\left(\sup_{R\in L_T} \gamma_T^{(1)}|Z(R)| \leq 1+2\epsilon\right) \\
&\qquad \geq \exp\left\{-C\frac{T}{a_T}\left(1+\log\frac{T}{a_T}\right)\left(1+\log\frac{b_T}{\sqrt{a_T}}\right)\right. \\
&\qquad\qquad \left. \times \exp\left\{-\frac{2(1+2\epsilon)^2}{2+\epsilon}(\log\Delta_T - \log\log\log T)\right\}\right\} \\
&\qquad = \exp\left\{-C\left(\frac{\Delta_T}{\log\log T}\right)^{-\epsilon}\log\log T\right\}.
\end{aligned}
$$

(4.2)

First, suppose that

$$\limsup_{T\to\infty} \frac{\log \Delta_T}{\log\log\log T} = +\infty.$$

There exists a sequence $T_N \uparrow +\infty$ such that

$$\frac{\log \Delta(T_N)}{\log\log\log T_N} \to +\infty.$$

From (4.2), it follows that for N large enough,

$$\mathsf{P}\left\{ \sup_{R\in L_{T_N}} \gamma^{(1)}(T_N)|Z(R)| \geq 1 + 2\epsilon \right\}$$

$$\leq C\left(\frac{\Delta(T_N)}{\log\log T_N}\right)^{-\epsilon} \log\log T_N$$

$$\leq C(\log\log T_N)^{-1} \to 0, (N \to \infty).$$

Then we have

$$\liminf_{T\to\infty} \sup_{R\in L_T} \gamma^{(1)}_T|Z(R)| \leq \liminf_{N\to\infty} \sup_{R\in L_{T_N}} \gamma^{(1)}(T_N)|Z(R)| \leq 1 + 2\epsilon \quad \text{a.s.,}$$

which implies (4.1). Now, we suppose that

$$(4.3) \qquad \frac{\log \Delta_T}{\log\log\log T} \leq r_1 < \infty, \text{ i.e., } \Delta_T \leq (\log\log T)^{r_1},$$

where $r_1 > 1$. It is easy to see that

$$(4.4) \qquad 1 \leq \liminf_{T\to\infty} \frac{\log\log b_T}{\log\log T} \leq \limsup_{T\to\infty} \frac{\log\log b_T}{\log\log T} < \infty.$$

Define

$$(4.5) \qquad D(T) = \begin{cases} \sup_{0\leq t\leq n+1} b(t), & \text{if } T = n, \\ \text{linear}, & \text{otherwise}. \end{cases}$$

Then

$$\sqrt{T} \leq b(T) \leq D(T) \leq \sup_{0\leq t\leq T+2} b(t) \leq c_0 b(T+2).$$

It follows that

$$(4.6) \qquad 1 \le \liminf_{T \to \infty} \frac{\log \log D(T)}{\log \log T} \le \limsup_{T \to \infty} \frac{\log \log D(T)}{\log \log T} < \infty.$$

Therefore, we can assume

$$(4.7) \qquad \frac{\log \Delta_T}{\log \log \log D(T)} \le r_0 < \infty,$$

where $r_0 > r_1$. On the other hand,

$$\log \frac{D(T)}{\sqrt{a(T)}} \le \log \frac{c_0 b(T+2)}{\sqrt{a(T+2)}} + \frac{1}{2} \log \frac{a(T+2)}{a(T)}$$

$$\le \log c + c(\log \log T)^{r_1} + \log \frac{T}{a_T}$$

$$\le c(\log \log T)^{r_1} \le c(\log \log b_T^2)^{r_1}$$

$$\le c(\log \log b_T)^{r_1} \le c(\log \log D(T))^{r_0},$$

i.e.,

$$(4.8) \qquad \log \frac{D(T)}{\sqrt{a(T)}} \le c(\log \log D(T))^{r_0}.$$

Now, from (4.6) it follows that

$$(4.9) \qquad \lim_{T \to \infty} \gamma_T^{(1)} \gamma_T^{*\,-1} = 1,$$

where $\gamma_T^* = \{2a_T^{2\alpha}(\log \Delta_T - \log \log \log D(T))\}^{-1/2}$. By the condition (iv) and (4.6), similarly to (4.2), we have

$$P\left\{ \sup_{R \in L_T} \gamma_T^* |Z(R)| \le 1 + 2\epsilon \right\}$$

$$(4.10) \qquad \ge \exp\left\{ -C\left(\frac{\Delta_T}{\log \log D(T)}\right)^{-\epsilon} \log \log D(T) \right\}$$

$$\ge c\big(\log D(T)\big)^{-1/4}.$$

Let $T_n = \sup\{T : D(T) \le e^{n^p}\}$. Then $T_n \uparrow \infty$. By the continuity of $D(T)$, we have $D(T_n) = e^{n^p}$. Define $d_n,\ A_n,\ X_n(\cdot,\cdot),\ \overline{X}_n(\cdot,\cdot),\ \overline{Y}_n(\cdot,\cdot)$ as in (3.11). Then $d_n = nD(T_n)$ and

$$\{\overline{X}_n(t,s); t,s \ge 0\} \overset{\mathcal{D}}{=} \left\{ D^{2\alpha}(T_n)\overline{Y}_n\left(\frac{t}{D(T_n)}, \frac{t}{D(T_n)}\right); t,s \ge 0 \right\}.$$

By noting that $b(T) \leq D(T)$, it follows from Lemma 3.1 that $E\overline{Y}_n^2(R) \leq \overline{\sigma}(|R|)$ for any $R \in L_{T_n}/D(T_n)$, where

$$
\begin{aligned}
L_{T_n}/D(T_n) = \{ R = [x_1, x_2] \times [y_1, y_2] : &\, 0 \leq x_1 \leq x_2 \leq b(T_n)/D(T_n), \\
&\, 0 \leq y_1 \leq y_2 \leq b(T_n)/D(T_n), x_2 y_2 \leq T_n/D^2(T_n), \\
&\, |R| \leq a(T_n)/D^2(T_n) \}.
\end{aligned}
$$

By the remark of Theoem 1.4 and (4.7), (4.8), we have

$$
\begin{aligned}
&P\left(\sup_{R \in L_{T_n}} \gamma^*(T_n)|\overline{X}_n(R)| \geq \epsilon \right) \\
&= P\left(\sup_{R \in L_{T_n}/D(T_n)} \gamma^*(T_n) D^{2\alpha}(T_n) |\overline{Y}_n(R)| \geq \epsilon \right) \\
&\leq C \frac{T_n}{a(T_n)} \left(1 + \log \frac{T_n}{a(T_n)} \right) \left(1 + \log \frac{b(T_n)}{\sqrt{a(T_n)}} \right) \\
&\quad \times \exp\left\{ - \frac{\epsilon^2 (\gamma^*(T_n))^{-2} D^{-4\alpha}(T_n)}{n^{-\delta} \left(\frac{a(T_n)}{D^2(T_n)} \right)^{2\alpha} \left(\log \frac{D^2(T_n)}{a(T_n)} \right)^{\gamma}} \right\} \\
&\leq C (\log\log D(T_n))^{r_0} \exp\left\{ -c \frac{\log(\Delta(T_n)/\log\log D(T_n))}{n^{-\delta}(\log n)^{\gamma p}} \right\} \\
&\leq C (\log n)^{r_0} \exp(-cn^{\delta/2}) \leq C \exp(-cn^{\delta/4}).
\end{aligned}
$$

(4.11)

From the Borel-Cantelli lemma it follows that

$$
(4.12) \qquad \limsup_{n \to \infty} \sup_{R \in L_{T_n}} \gamma^*(T_n)|\overline{X}_n(R)| \leq \epsilon \quad \text{a.s.}
$$

On the other hand, from (4.10) and (4.11) it follows that

$$
\begin{aligned}
&\sum_{n=1}^{\infty} P\left\{ \sup_{R \in L_{T_n}} \gamma^*(T_n)|X_n(R)| \leq 1 + 3\epsilon \right\} \\
&\geq \sum_{n=1}^{\infty} \left\{ P\left(\sup_{R \in L_{T_n}} \gamma^*(T_n)|Z(R)| \leq 1 + 2\epsilon \right) - P\left(\sup_{R \in L_{T_n}} \gamma^*(T_n)|\overline{X}_n(R)| \leq \epsilon \right) \right\} \\
&\geq \sum_{n=1}^{\infty} \left\{ n^{-p/4} - C \exp(-cn^{\delta/4}) \right\} = \infty.
\end{aligned}
$$

It follows from the Borel-Cantelli lemma and the independence of $\{X_n(\cdot, \cdot)\}$ that

$$
(4.13) \qquad \liminf_{n \to \infty} \sup_{R \in L_{T_n}} \gamma^*(T_n)|X_n(R)| \leq 1 + 3\epsilon \quad \text{a.s.}
$$

Combining (4.9), (4.12) and (4.13) yields

$$
\begin{aligned}
\liminf_{T\to\infty} \gamma_T^{(1)} &\sup_{R\in L_T} |Z(R)| \\
&\le \liminf_{n\to\infty} \sup_{R\in L_{T_n}} \gamma^*(T_n)|Z(R)| \\
&\le \liminf_{n\to\infty} \sup_{R\in L_{T_n}} \gamma^*(T_n)|X_n(R)| + \limsup_{n\to\infty} \sup_{R\in L_{T_n}} \gamma^*(T_n)|\overline{X}_n(R)| \\
&\le 1 + 4\epsilon \quad \text{a.s.}
\end{aligned}
$$

This completes the proof.

Proposition 4.2. *Let $0 < a_T \le T$ and $b_T \ge \sqrt{T}$ be two functions of T such that b_T is quasi-increasing. Let $\overline{\Delta}_T = Ta_T^{-1}(1 + \log(b_T T^{-1/2}))$. Assume that*

$$
\tag{4.14}
\liminf_{T\to\infty} \frac{\log \overline{\Delta}_T}{\log\log\log T} = r > 1.
$$

If $r = +\infty$ or $0 < \alpha \le 1/2$ or $\liminf_{T\to\infty} a_T/T > 0$, then

$$
\tag{4.15}
\liminf_{T\to\infty} \sup_{R\in L_T^*} \gamma_T^{(2)} Z(R) \ge 1 \quad \text{a.s.,}
$$

where $\gamma_T^{(2)} = \{2a_T^{2\alpha}(\log \overline{\Delta}_T - \log\log\log T)\}^{-1/2}$.

The proof of this proposition will be given by three steps:

Step 1. *Let $0 < a_T \le T$, $b_T \ge \sqrt{T}$ be of functions of T such that b_T is quasi-increasing. We have*

$$
\tag{4.16}
\liminf_{T\to\infty} \inf_{0<a_T\le\sqrt{T}} \sup_{R\in L_T^*} \gamma_T^{(2)} Z(R) \ge 1 \quad \text{a.s.,}
$$

where $\inf_{0<a_T\le\sqrt{T}}$ means $\inf_{\{T:0<a_T\le\sqrt{T}\}}$.

Proof. Let $a_T^* = a_T \wedge \sqrt{T}$. Define $L_T^*(a_T^*)$, $\gamma_T^{(2)}(a_T^*)$ similar to L_T^*, $\gamma_T^{(2)}$ with a_T^* taking the place of a_T. Then

$$
\begin{aligned}
\liminf_{T\to\infty} \inf_{0<a_T\le\sqrt{T}} \sup_{R\in L_T^*} \gamma_T^{(2)} Z(R) &= \liminf_{T\to\infty} \inf_{0<a_T\le\sqrt{T}} \sup_{R\in L_T^*(a_T^*)} \gamma_T^{(2)}(a_T^*)Z(R) \\
&\ge \liminf_{T\to\infty} \sup_{R\in L_T^*(a_T^*)} \gamma_T^{(2)}(a_T^*)Z(R).
\end{aligned}
$$

So, without loss of generality we can assume that $0 < a_T \leq \sqrt{T}$ for all T. Then

$$(4.17) \qquad \frac{Ta_T^{-1}}{\log\log T} \to \infty, \qquad \frac{\log C_T}{\log\log(a_T^\alpha + a_T^{-\alpha})} \to \infty, \qquad \gamma_T^{(2)}\overline{\delta}_T^{-1} \to 1,$$

where C_T and $\overline{\delta}_T$ is defined as in Step 1 of the proof of Theorem 1.1. Let $1 < \theta < 2$. Define

$$
\begin{cases}
A_k = \{T : \theta^k \leq a_T^\alpha < \theta^{k+1}\}, & -\infty < k < \infty, \\
B_i = \{T : 2^i \leq C_T < 2^{i+1}\}, & i \geq 0, \\
T_{k,i} = \sup\{T : T \in A_k B_i\}, & T_{k,i}^* = \inf\{T : T \in A_k B_i\}, \\
b_{k,i} = \sup\{b(T) : T \in A_k b_i\}, & b_{k,i}^* = \inf\{b(T) : T \in A_k b_i\}, \\
L_{k,i}^* = \big\{R = [x_1,x_2] \times [y_1,y_2] : 0 \leq x_1 \leq x_2 \leq b_{k,i}^*, \\
\qquad 0 \leq y_1 \leq y_2 \leq b_{k,i}^*, x_2 y_2 \leq T_{k,i}^*, |R| = \theta^k\big\} \\
L_{k,i}' = \big\{R = [x_1,x_2] \times [y_1,y_2] : 0 \leq x_1 \leq x_2 \leq b_{k,i}, \\
\qquad 0 \leq y_1 \leq y_2 \leq b_{k,i}, x_2 y_2 \leq T_{k,i}, |R| \leq \theta^k(\theta-1)\big\}.
\end{cases}
$$

Then we have

$$\liminf_{T\to\infty} \sup_{R\in L_T^*} \gamma_T^{(2)} Z(R) = \liminf_{T\to\infty} \sup_{R\in L_T^*} \overline{\delta}_T Z(R)$$

$$\geq \liminf_{|k|+l\to\infty} \inf_{i\geq l} \sup_{T\in A_k B_i} \sup_{R\in L_T^*} \overline{\delta}_T Z(R)$$

$$(4.18) \qquad \geq \liminf_{\substack{|k|+l\to\infty \\ A_k B_i \neq \emptyset}} \inf_{i\geq l} \sup_{R\in L_{k,i}^*} Z(R)\big/\big\{2\theta^{2(k+1)}\big(\log 2^{i+1} + \log\log\theta^{|k|+1}\big)\big\}^{1/2}$$

$$- 4\limsup_{\substack{|k|+l\to\infty \\ A_k B_i \neq \emptyset}} \sup_{i\geq l} \sup_{R\in L_{k,i}'} |Z(R)|\big/\big\{2\theta^{2k}\big(\log 2^i + \log\log\theta^{|k|}\big)\big\}^{1/2}$$

$$=: I_1 - 4I_2.$$

Similarly to (3.9), we have

$$I_2 \leq (\theta-1)\limsup_{\substack{|k|+l\to\infty \\ A_k B_i \neq \emptyset}} \sup_{i\geq l} \sup_{R\in L_{k,i}'}$$

$$(4.19) \qquad\qquad |Z(R)|\big/\big\{2(\theta^k(\theta-1))^2\big(\log 2^i + \log\log\theta^{|k|}\big)\big\}^{1/2}$$

$$\leq (\theta-1)\theta^2 \quad \text{a.s.}$$

Now, we consider I_1. By Theorem 1.3, we have, for i, k large enough,

$$P\left(\sup_{R\in L_{k,i}^*} Z(R)/\{2\theta^{2(k+1)}(\log 2^{i+1} + \log\log \theta^{|k|+1})\}^{1/2} \leq 1/\theta^2\right)$$

$$\leq \exp\left\{-C\frac{T_{k,i}^*}{\theta^k}\left(1 + \log\frac{b_{k,i}^*}{\sqrt{T_{k,i}}}\right)\exp\left\{-\frac{1}{\theta}(\log 2^{i+1} + \log\log\theta^{|k|+1})\right\}\right\}$$

$$+ \exp\left\{-2(\log 2^{i+1} + \log\log\theta^{|k|+1})\right\}.$$

Similarly to (3.7), we have

$$\frac{T_{k,i}^*}{\theta^k}\left(1 + \log\frac{T_{k,i}^*}{\theta^k}\right)\left(1 + \log\frac{b_{k,i}^*}{\sqrt{\theta^k}}\right) \geq C2^i.$$

It follows that

$$\frac{T_{k,i}^*}{\theta^k}\left(1 + \log\frac{b_{k,i}^*}{\sqrt{T_{k,i}}}\right) \geq \left(\frac{T_{k,i}^*}{\theta^k}\left(1 + \log\frac{T_{k,i}^*}{\theta^k}\right)\left(1 + \log\frac{b_{k,i}^*}{\sqrt{\theta^k}}\right)\right)^{1/\sqrt{\theta}}$$

$$\geq C2^{i/\sqrt{\theta}}.$$

It follows that

$$P\left(\sup_{R\in L_{k,i}^*} Z(R)/\{2\theta^{2(k+1)}(\log 2^{i+1} + \log\log \theta^{|k|+1})\}^{1/2} \leq 1/\theta^2\right)$$

$$\leq \exp\left(-C2^{i/\sqrt{\theta}}(2^i|k|)^{-1/\theta}\right) + (2^i\log|k|)^{-2}$$

$$\leq \exp\left(-C2^{i\epsilon}|k|^{-1/\theta}\right) + 2^{-i}|k|^{-2}$$

$$\leq \exp\left(-C2^{i\epsilon/2}|k|^{\epsilon}\right) + 2^{-i}|k|^{-2} \leq \exp\left(-C2^{i\epsilon/2} - |k|^{\epsilon}\right) + 2^{-i}|k|^{-2},$$

where the last two inequalities used the facts that $ab \geq a + b$ $(a, b \geq 2)$ and

$$\frac{\log 2^{i+1}}{\log\log\theta^{|k|}} \geq \frac{\log C_T}{\log\log(a_T^{\alpha} + a_T^{-\alpha})} \to \infty.$$

It follows that

$$\sum_{l=1}^{\infty}\sum_{|k|=1}^{\infty} P\left(\inf_{\substack{i\geq l \\ A_k b_i \neq \emptyset}}\sup_{R\in L_{k,i}^*} Z(R)/\{2\theta^{2(k+1)}(\log 2^{i+1} + \log\log\theta^{|k|+1})\}^{1/2} \leq 1/\theta^2\right)$$

$$\leq \sum_{l=1}^{\infty}\sum_{|k|=1}^{\infty}\sum_{i=l}^{\infty}\left\{\exp(-C2^{i\epsilon/2})\exp(-|k|^{\epsilon}) + 2^{-i}|k|^{-2}\right\} < +\infty,$$

which together with the Borel-Cantelli lemma implies

$$(4.20)\qquad\qquad\qquad\qquad I_1 \geq \frac{1}{\theta^2}\quad\text{a.s.}$$

Putting (4.18), (4.19), (4.20) together and letting $\theta \to 1$ yield (4.16).

Step 2. *Let $0 < a_T \leq T$, $b_T \geq \sqrt{T}$ be of functions of T such that b_T is quasi-increasing. If*

$$(4.21) \qquad \lim_{T \to \infty} \frac{\log \overline{\Delta}_T}{\log \log \log T} = +\infty,$$

then

$$(4.22) \qquad \liminf_{T \to \infty} \gamma_T^{(2)} \sup_{R \in L_T^*} Z(R) \geq 1 \quad \text{a.s.}$$

Proof. We have

$$\liminf_{T \to \infty} \gamma_T^{(2)} \sup_{R \in L_T^*} Z(R)$$

$$\geq \min \left\{ \liminf_{T \to \infty} \inf_{0 < a_T \leq \sqrt{T}} \gamma_T^{(2)} \sup_{R \in L_T^*} Z(R), \liminf_{T \to \infty} \inf_{\sqrt{T} \leq a_T} \gamma_T^{(2)} \sup_{R \in L_T^*} Z(R) \right\}.$$

By Step 1, it is sufficient to show that

$$\liminf_{T \to \infty} \inf_{\sqrt{T} \leq a_T} \gamma_T^{(2)} \sup_{R \in L_T^*} Z(R) \geq 1 \quad \text{a.s.}$$

Also, without loss of generality, we can assume that $a_T \geq \sqrt{T}$ for all T. Then

$$\lim_{T \to \infty} \frac{\log \log a_T}{\log \log T} = 1, \qquad \lim_{T \to \infty} \frac{\log \overline{\Delta}_T}{\log \log \log a_T} = +\infty.$$

Also, (4.21) implies that $\overline{\Delta}_T / \log \log T \to \infty$. Let

$$(4.23) \qquad \begin{cases} A_k = \{T : e^{\sqrt{k}} \leq a_T^\alpha < e^{\sqrt{k+1}}\}; \quad k \geq 1, \\[4pt] B_i = \{T : 2^i \leq \overline{\Delta}_T < 2^{i+1}\}, \qquad i \geq 0, \\[4pt] T_{k,i} = \sup\{T : T \in A_k B_i\}, \qquad T_{k,i}^* = \inf\{T : T \in A_k B_i\}, \\[4pt] b_{k,i} = \sup\{b(T) : T \in A_k b_i\}, \quad b_{k,i}^* = \inf\{b(T) : T \in A_k b_i\}, \\[4pt] L_{k,i}^* = \big\{ R = [x_1, x_2] \times [y_1, y_2] : 0 \leq x_1 \leq x_2 \leq b_{k,i}^*, \\[4pt] \qquad\qquad 0 \leq y_1 \leq y_2 \leq b_{k,i}^*, x_2 y_2 \leq T_{k,i}^*, |R| = e^{\sqrt{k}} \big\}, \\[4pt] L_{k,i}' = \big\{ R = [x_1, x_2] \times [y_1, y_2] : 0 \leq x_1 \leq x_2 \leq b_{k,i}, \\[4pt] \qquad\qquad 0 \leq y_1 \leq y_2 \leq b_{k,i}, x_2 y_2 \leq T_{k,i}, |R| \leq e^{\sqrt{k+1}} - e^{\sqrt{k}} \big\}. \end{cases}$$

Then we have

$$\liminf_{T\to\infty} \sup_{R\in L_T^*} \gamma_T^{(2)} Z(R)$$

$$= \liminf_{T\to\infty} \sup_{R\in L_T^*} Z(R)/\{2a_T^{2\alpha}(\log\overline{\Delta}_T - \log\log\log a_T)^{1/2}$$

$$\text{(4.24)}\qquad \geq \liminf_{\substack{l+k\to\infty \\ A_k B_i\neq\emptyset}} \inf_{i\geq k} \sup_{R\in L_{k,i}^*} Z(R)/\{2e^{2\sqrt{k+1}}(\log 2^{i+1} - \log\log\log e^{\sqrt{k}})\}^{1/2}$$

$$-4\limsup_{\substack{l+k\to\infty \\ A_k B_i\neq\emptyset}} \sup_{i\geq k} \sup_{R\in L_{k,i}'}$$

$$|Z(R)|/\{2e^{2\sqrt{k}}(\log 2^i - \log\log\log e^{\sqrt{k+1}})\}^{1/2}$$

$$=: I_1 - 4I_2.$$

Simialrly to (3.7), one can show that

$$\frac{T_{k,i}}{e^{\sqrt{k+1}}}\left(1 + \log\frac{T_{k,i}}{e^{\sqrt{k+1}}}\right)\left(1 + \log\frac{b_{k,i}}{e^{\sqrt{k+1}/2}}\right) \leq \left\{\frac{T_{k,i}}{e^{\sqrt{k+1}}}\left(1 + \log\frac{b_{k,i}}{\sqrt{T_{k,i}}}\right)\right\}^2$$

$$\leq C\left(\sup_{T\in A_k B_i}\overline{\Delta}_T\right)^2 \leq C2^{2i},$$

$$\frac{T_{k,i}^*}{e^{\sqrt{k+1}}}\left(1 + \log\frac{b_{k,i}^*}{\sqrt{T_{k,i}^*}}\right) \geq C\inf_{T\in A_k B_i}\overline{\Delta}_T \geq C2^i.$$

Note that $e^{\sqrt{k+1}} - e^{\sqrt{k}} \approx e^{\sqrt{k}}/\sqrt{k}$, and $2^i/\log\log e^{\sqrt{k+1}} \geq C\overline{\Delta}_T/\log\log T \to \infty$ along $A_k B_i$. It follows from the remark of Theorem 1.4 that

$$\mathsf{P}\left\{\sup_{R\in L_{k,i}'} |Z(R)|/\{2e^{\sqrt{k}}(\log 2^i - \log\log\log e^{\sqrt{k+1}})\}^{1/2} \geq \epsilon\right\}$$

$$\leq C\frac{T_{k,i}}{e^{\sqrt{k+1}} - e^{\sqrt{k}}}\left(1 + \log\frac{T_{k,i}}{e^{\sqrt{k+1}} - e^{\sqrt{k}}}\right)\left(1 + \log\frac{b_{k,i}}{\sqrt{e^{\sqrt{k+1}} - e^{\sqrt{k}}}}\right)$$

$$\times \exp\left\{-\frac{2\epsilon^2 e^{2\sqrt{k}}(\log 2^i - \log\log\log e^{\sqrt{k+1}})}{(2+\epsilon)(e^{\sqrt{k+1}} - e^{\sqrt{k}})^2}\right\}$$

$$\leq C\sqrt{k}\log k2^{2i}\exp\left\{-Ck(\log 2^i - \log\log k)\right\}$$

$$\leq C\sqrt{k}\log k2^{2i}\exp\left\{-4(\log 2^i - \log\log k)\right\}\exp(-k/4)$$

$$\leq c\sqrt{k}\log k2^{-2i}(\log k)^4 \exp(-k/4) \leq C2^{-2i}\exp(-k/8)$$

for k large enough, where the inequality $ab \geq a + b$ $(a, b \geq 2)$ is used. Since

$$\sum_{l=1}^{\infty}\sum_{k=1}^{\infty}\sum_{i=l}^{\infty} 2^{-2i}\exp(-k/8) < \infty,$$

from the Borel-Cantelli lemma, it follows that

$$(4.25) \qquad\qquad I_2 = 0 \quad \text{a.s.}$$

On the other hand, by (1.10) in Theorem 1.4, we have

$$\mathrm{P}\left\{ \sup_{R \in L_{k,i}^*} Z(R)/\{2e^{\sqrt{k+1}}(\log 2^{i+1} - \log\log\log e^{\sqrt{k}})\}^{1/2} \leq 1 - 2\epsilon \right\}$$

$$\leq \exp\left\{ -C\frac{1}{m}\frac{T_{k,i}^*}{e^{\sqrt{k}}}\left(1 + \log\frac{b_{k,i}^*}{\sqrt{T_{k,i}^*}}\right) \right.$$

$$(4.26) \qquad \times \exp\left\{ -\frac{2(1-2\epsilon)^2 e^{2\sqrt{k+1}}(\log 2^i - \log\log\log e^{\sqrt{k}})}{(2-\epsilon)e^{2\sqrt{k}}} \right\}\right\}$$

$$+ \exp\left\{ -Cm^{2-2\alpha}(\log 2^i - \log\log\log e^{\sqrt{k}}) \right\}$$

$$\leq \exp\left\{ -C\frac{1}{m}\left(\frac{2^i}{\log k}\right)^{\epsilon}\log k \right\} + \exp\left\{ -Cm^{2-2\alpha}\log\frac{2^i}{\log k} \right\}$$

$$=: J_{i,k}(1) + J_{i,k}(2).$$

Let $m = (2^i/\log k)^{\epsilon/2}$. Then, from the fact $2^i/\log k \to \infty$ along $A_k B_i$, it follows that

$$\sum_{l=1}^{\infty}\sum_{k=1}^{\infty}\sum_{i=l}^{\infty} J_{i,k}(1) \leq \sum_{l=1}^{\infty}\sum_{k=1}^{\infty}\sum_{i=l}^{\infty}\exp\left\{ -\frac{C}{2}\left(\frac{2^i}{\log k}\right)^{\epsilon/2} \right\}k^{-2}$$

$$(4.27) \qquad \leq C\sum_{l=1}^{\infty}\sum_{k=1}^{\infty}\sum_{i=l}^{\infty}\left(\frac{2^i}{\log k}\right)^{-2}k^{-2}$$

$$\leq \sum_{l=1}^{\infty}\sum_{k=1}^{\infty}\sum_{i=l}^{\infty} 2^{-2i}k^{-2}(\log k)^2 < \infty.$$

Since $(\log \overline{\Delta}_T)/\log\log\log a_T \to \infty$, it follows that, for i, k large enough,

$$m^{2-2\alpha} \geq \left(\frac{2^i}{\log k}\right)^{\epsilon(2-2\alpha)/4}\log k.$$

Then, it is similar to (4.27) that

$$(4.29) \quad \sum_{l=1}^{\infty}\sum_{k=1}^{\infty}\sum_{i=l}^{\infty} J_{i,k}(2) \le \sum_{l=1}^{\infty}\sum_{k=1}^{\infty}\sum_{i=l}^{\infty} \exp\left\{ -C\left(\frac{2^i}{\log k}\right)^{\epsilon(2-2\alpha)/4} \log k \right\}$$
$$< +\infty.$$

Combining (4.26), (4.27) and (4.29) yields

$$(4.30) \qquad\qquad\qquad I_1 \ge 1 \quad \text{a.s.}$$

From (4.25), (4.30) and (4.24), it follows that (4.22) is true.

Step 3. Let $0 < a_T \le T$, $b_T \ge \sqrt{T}$ be of functions of T such that b_T is *quasi-increasing. Assume that*

$$(4.31) \qquad\qquad \liminf_{T\to\infty} \frac{\log \overline{\Delta}_T}{\log\log\log T} = r > 1.$$

If $0 < \alpha \le 1/2$ or $\liminf_{T\to\infty} a_T/T > 0$, then

$$(4.32) \qquad\qquad \liminf_{T\to\infty} \gamma_T^{(2)} \sup_{R\in L_T^*} Z(R) \ge 1 \quad \text{a.s.}$$

Proof. Obviously, (4.31) implies that $\overline{\Delta}_T/\log\log T \to \infty$. Note that in the proof of Step 2, the condition (4.21) was only used in showing that the sum of $J_{k,i}(2)$ in (4.26) is finite. Now, using the inequalities (1.11) and (1.12) in Theorem 1.3 instead of (1.10), we can get

$$J_{k,i}(2) \le \exp\left\{ -Ce^{(2-2\alpha)m} \log \frac{2^i}{\log k} \right\},$$

where $m = (2^i/\log k)^{\epsilon/2}$. By the condition (4.31), we have

$$\frac{2^i}{\log k} \ge c\frac{\overline{\Delta}_T}{\log\log a_T} \ge c(\log\log a_T)^{(r-1)/2} \ge c(\log k)^{(r-1)/2}$$

for i, k large enough and $T \in A_k B_i$. It follows that

$$e^{(2-2\alpha)m} \log \frac{2^i}{\log k} \ge cm^{2M/\epsilon} \ge C\frac{2^i}{\log k} \log k,$$

where $M \geq 4/(r-1) + 1$. It follows that

$$J_{k,i}(2) \leq \exp\left\{ -C\frac{2^i}{\log k}\log k \right\} \leq \exp\left\{ -\frac{C}{2}\frac{2^i}{\log k} \right\}k^{-2}.$$

Then, it is similar to (4.27) that

$$\sum_{l=1}^{\infty}\sum_{k=1}^{\infty}\sum_{i=l}^{\infty} J_{i,k}(2) < \infty.$$

The rest proofs are similar to that of Step 2.

Finally, if the condition (iii') holds, then

$$\frac{\log \overline{\Delta}_T}{\log Ta_T^{-1} + \log(1 + \log b_T/\sqrt{T})} \to 1, \quad \gamma_T \sim \gamma_T^{(1)} \sim \gamma_T^{(2)} \quad (T \to \infty).$$

Combining Propositions 4.1 and 4.2 completes the proof of Theorem 1.2. This completes the proof.

REFERENCES

1. Y. K. Choi and N. Kôno, *How big are the increments of a two-parameter Gaussian process?*, J. Theor. Probab. **12** (1999), 105–129.

2. E. Csáki, M. Csörgő and Q. M. Shao, *Fernique type inequalities and moduli of continuity for l^2-valued Ornstein-Uhlenbeck processes*, Ann. Inst. Henri. Poincaré Probab. Statist. **28** (1992), 479–517.

3. M. Csörgő and Q. M. Shao, *Strong limit theorems for large and small increments of l^p-valued Gaussian processes*, Ann. Probab. **21** (1993), 1958–1990.

4. M. Csörgő and P. Révész, *Strong Approximations in probability and statistics*, New York, Academic Press, 1981.

5. C. G. Khatri, *On certain inequalities for normal distributions and theiplications to simultaneous confidence bounds*, Ann. Math. Stat. **38** (1967), 1853–1867.

6. F. C. Kong, *The increments of a two-parameter Gaussian process*, J. of Anuhui University **10** (1989), 8–18; (in Chinese).

7. Z. Y. Lin and Y. K. Choi, *Some limit theorem for fractional Lévy-Brownian fields*, Stochastic Processes Their Appl. **82** (1999), 229–244.

8. D. Monrad and H. Rootzen, *Small values of Gaussian processes and functional laws of the iterated lagarithm*, Probab. Theory Rel. Fields **101** (1995), 173–192.

9. J. Ortega, *On the size of the increments of non-stationary Gaussian processes*, Stochastic Processes Their Appl. **18** (1984), 47–56.

10. Z. Šidák, *On multivariate normal probabilities of rectangles: their dependence on correlation*, Ann. Math. Stat. **39** (1968), 1425–1434.

11. D. Slepian, D, *The one-sided barrier problem for Gaussian noise*, Bell System Tech. J. **41** (1962), 463–501.

12. L. X. Zhang, *Some liminf results on increments of fractional Brownian motion*, Acta Math. Hungar. **71** (1996a), 209–234.

13. L. X. Zhang, *Two different kinds of liminf results on the LIL of the two-parameter Wiener processes*, Stochastic Processes Their Appl. **71** (1996b), 215–240.
14. L. X. Zhang, *A note on liminfs for increments of a fractional Brownain mation*, Acta Math. Hungar. **76** (1997a), 145–154.
15. L. X. Zhang, *A liminf of the increments of two-parameter Wiener process*, Chinese Ann. Math. **18** (1997b), 235–246; (in Chinese); Chinese J. of Contemporary Math. **18**, 185–198.

INDEX